International Association of Fire Chiefs

Fundamentals of Fire Fighter Skills

Evidence-Based Practices

ENHANCED THIRD EDITION

Student Workbook

JONES & BARTLETT LEARNING

Jones & Bartlett Learning
World Headquarters
5 Wall Street
Burlington, MA 01803
978-443-5000
info@jblearning.com
www.jblearning.com

National Fire Protection Association
1 Batterymarch Park
Quincy, MA 02169-7471
www.NFPA.org

International Association of Fire Chiefs
4025 Fair Ridge Drive
Fairfax, VA 2
www.IAFC.org

Jones & Bartlett Learning books and products are available through most bookstores and online booksellers. To contact Jones & Bartlett Learning directly, call 800-832-0034, fax 978-443-8000, or visit our website, www.jblearning.com.

> Substantial discounts on bulk quantities of Jones & Bartlett Learning publications are available to corporations, professional associations, and other qualified organizations. For details and specific discount information, contact the special sales department at Jones & Bartlett Learning via the above contact information or send an email to specialsales@jblearning.com.

Editorial Credits
Author: Douglas C. Ott

Production Credits
General Manager, Vocational Solutions: Doug Kaplan
Vice President, Publisher: Kimberly Brophy
Vice President of Sales, Public Safety Group: Matthew Maniscalco
Director of Sales, Public Safety Group: Patricia Einstein
Executive Editor: William Larkin
Senior Development Editor: Janet Morris
Associate Managing Editor: Amanda Brandt
Associate Production Editor: Kristen Rogers
Senior Marketing Manager: Brian Rooney
VP, Manufacturing and Inventory Control: Therese Connell
Composition: diacriTech
Cover Design: Kristin E. Parker
Director of Photo Research and Permissions: Amy Wrynn
Rights & Photo Research Assistant: Ashley Dos Santos
Cover Image: © Photo by David Traiforos, luvfireimages.com
Printing and Binding: Edwards Brothers Malloy
Cover Printing: Edwards Brothers Malloy

Copyright © 2017 by Jones & Bartlett Learning, LLC, an Ascend Learning Company, and the National Fire Protection Association®

All rights reserved. No part of the material protected by this copyright may be reproduced or utilized in any form, electronic or mechanical, including photocopying, recording, or by any information storage and retrieval system, without written permission from the copyright owner.

The content, statements, views, and opinions herein are the sole expression of the respective authors and not that of Jones & Bartlett Learning, LLC. Reference herein to any specific commercial product, process, or service by trade name, trademark, manufacturer, or otherwise does not constitute or imply its endorsement or recommendation by Jones & Bartlett Learning, LLC and such reference shall not be used for advertising or product endorsement purposes. All trademarks displayed are the trademarks of the parties noted herein. *Fundamentals of Fire Fighter Skills: Evidence-Based Practices, Enhanced Third Edition, Student Workbook* is an independent publication and has not been authorized, sponsored, or otherwise approved by the owners of the trademarks or service marks referenced in this product.

There may be images in this book that feature models; these models do not necessarily endorse, represent, or participate in the activities represented in the images. Any screenshots in this product are for educational and instructive purposes only. Any individuals and scenarios featured in the case studies throughout this product may be real or fictitious, but are used for instructional purposes only.

The procedures and protocols in this book are based on the most current recommendations of responsible medical sources. The International Association of Fire Chiefs (IAFC), National Fire Protection Association (NFPA®), and the publisher, however, make no guarantee as to, and assume no responsibility for, the correctness, sufficiency, or completeness of such information or recommendations. Other or additional safety measures may be required under particular circumstances.

ISBN: 978-1-284-23962-1

6048

Printed in the United States of America
19 18 17 16 15   10 9 8 7 6 5 4 3 2 1

Table of Contents

> Notes to the student: Exercises indicated with a 🎩 are specific to the Fire Fighter II level. Consult your instructor for access to the Student Workbook Answer Key.

Chapter 1 2
The Orientation and History of the Fire Service

Chapter 2 10
Fire Fighter Safety

Chapter 3 18
Personal Protective Equipment and Self-Contained Breathing Apparatus

Chapter 4 28
Fire Service Communications

Chapter 5 38
Incident Command System

Chapter 6 48
Fire Behavior

Chapter 7 58
Building Construction

Chapter 8 66
Portable Fire Extinguishers

Chapter 9 78
Fire Fighter Tools and Equipment

Chapter 10 88
Ropes and Knots

Chapter 11 102
Response and Size-Up

Chapter 12 112
Forcible Entry

Chapter 13 124
Ladders

Chapter 14 138
Search and Rescue

Chapter 15 150
Ventilation

Chapter 16 166
Water Supply

Chapter 17 180
Fire Attack and Foam

Chapter 18 192
Fire Fighter Survival

Chapter 19 202
Salvage and Overhaul

Chapter 20 214
Fire Fighter Rehabilitation

Chapter 21 222
Wildland and Ground Fires

Chapter 22 230
Fire Suppression

Chapter 23 244
Preincident Planning

Chapter 24 256
Fire and Emergency Medical Care

Chapter 25 266
Emergency Medical Care

Chapter 26 280
Vehicle Rescue and Extrication

Chapter 27 292
Assisting Special Rescue Teams

Chapter 28 302
Hazardous Materials: Overview

Chapter 29 310
Hazardous Materials: Properties and Effects

Chapter 30 322
Hazardous Materials: Recognizing and Identifying the Hazards

Chapter 31 332
Hazardous Materials: Implementing a Response

Chapter 32 340
Hazardous Materials: Personal Protective Equipment, Scene Safety, and Scene Control

Chapter 33 350
Hazardous Materials: Response Priorities and Actions

Chapter 34 360
Hazardous Materials: Decontamination Techniques

Chapter 35 368
Terrorism Awareness

Chapter 36 378
Fire Prevention and Public Education

Chapter 37 388
Fire Detection, Protection and Suppression Systems

Chapter 38 398
Fire Cause Determination

Join Our Growing Online Community Today!

Connect with other students in the public safety field, keep up with relevant news and events, receive special offers, and much more!

 Facebook:
www.fb.com/JBLFireandEMS

 Blog:
blogs.jblearning.com/public-safety

 Facebook:
@JBL_EMS_Fire

 25% Off Your First Purchase When You Sign Up for Email Alerts at:
www.jblearning.com/eUpdates

The Orientation and History of the Fire Service

Workbook Activities

The following activities have been designed to help you. Your instructor may require you to complete some or all of these activities as a regular part of your fire fighter training program. You are encouraged to complete any activity that your instructor does not assign as a way to enhance your learning in the classroom.

Chapter Review

The following exercises provide an opportunity to refresh your knowledge of this chapter.

Matching

Match each of the terms in the left column to the appropriate definition in the right column.

_____ 1. Fireplug
_____ 2. Doff
_____ 3. SCBA
_____ 4. Battalion chief
_____ 5. Fire hook
_____ 6. Don
_____ 7. Safety officer
_____ 8. Captain
_____ 9. Driver/operator
_____ 10. Discipline

A. The position responsible for a fire company and for coordinating activities of that company among the shifts
B. A valve installed to control water accessed from wooden pipes
C. Guiding and directing fire fighters to do what their fire department expects of them
D. To take off
E. The position responsible for operating the fire apparatus
F. A tool used to pull down burning structures
G. To put on
H. The position often in charge of running calls and supervising multiple stations or districts within a city
I. The position with the authority to stop any firefighting activity until it can be done safely and correctly
J. Respirator with independent air supply; air packs used by fire fighters

Multiple Choice

Read each item carefully, and then select the best response.

_____ 1. To provide a uniform way to deal with emergency situations, departments develop and follow
   A. laws.
   B. regulations.
   C. standard operating procedures (SOPs).
   D. policies.

_____ 2. In some fire departments, the preferred terminology for standard operating procedures is
   A. policies.
   B. regulations.
   C. rules.
   D. suggested operating guidelines.

CHAPTER 1

_____ 3. The majority of fire departments consist of
   A. all career fire fighters.
   B. mostly career fire fighters.
   C. mostly volunteer fire fighters.
   D. all volunteer fire fighters.

_____ 4. The organizational structure of a fire department consists of a(n)
   A. chain of custody.
   B. incident management system.
   C. chain of command.
   D. division of labor.

_____ 5. New fire fighters usually report to a
   A. lieutenant.
   B. captain.
   C. battalion chief.
   D. division chief.

_____ 6. Most experts believe that span of control should not exceed how many people in a complex or rapidly changing environment?
   A. 3
   B. 4
   C. 5
   D. 6

_____ 7. In the 1700s, a fire mark indicated
   A. the homeowner had fire insurance.
   B. the homeowner was a career fire fighter.
   C. the homeowner was a volunteer fire fighter.
   D. the home had a previous fire.

_____ 8. The overall responsibility for the administration and operations of the department belongs to the
   A. battalion chief.
   B. chief of the department.
   C. incident commander.
   D. government.

_____ 9. The first water system valves or fire hydrants used by fire fighters were called
   A. fire taps.
   B. water valves.
   C. water boxes.
   D. fireplugs.

_____ 10. The theory that each fire fighter answers to only one supervisor is referred to as
   A. unity of command.
   B. span of control.
   C. division of labor.
   D. discipline.

FUNDAMENTALS OF FIRE FIGHTER SKILLS

_____ 11. Which of the following is not a form of discipline?
   A. SOPs
   B. Policies
   C. Span of control
   D. Training

_____ 12. Augustus Caesar created what was probably the first fire department, called the Familia Publica, in
   A. 100 B.C.
   B. 24 B.C.
   C. 1 B.C.
   D. 10 A.D.

_____ 13. What are the most important "machines" on the fire scene?
   A. Hand tools
   B. Engines
   C. Well-trained fire fighters
   D. Ladders

_____ 14. The first fire insurance company in the United States was established in 1736
   A. by George Washington.
   B. by Benjamin Franklin.
   C. in Charleston, South Carolina.
   D. by the Alexandria Fire Department.

_____ 15. Colonial fire fighters had limited equipment; most departments had only buckets, ladders, and
   A. hand-powered pumpers.
   B. horse-drawn water carriages.
   C. fire hooks.
   D. hoses.

_____ 16. The company responsible for securing a water source, deploying handlines, and putting water on the fire is the
   A. truck company.
   B. brush company.
   C. water company.
   D. engine company.

_____ 17. Before radios or bullhorns, the _____ allowed communication during an incident. Today it serves as a symbol of authority.
   A. chief's trumpet
   B. call box
   C. monitor
   D. commander's horn

_____ 18. The company that specializes in forcible entry, ventilation, roof operations, search and rescue, and ground ladders is the
   A. truck company.
   B. brush company.
   C. water company.
   D. engine company.

_____ 19. In 1871, a historic fire, which was believed to have been started by a cow, burned for three days, destroyed more than 2000 acres and 17,000 homes, and killed 300 people. This was the
   A. Great Chicago Fire.
   B. Peshtigo Fire.
   C. Green Bay Burn.
   D. Alexandria Fire.

_____ 20. The fire service draws its authority from the governing entity, and the head of the department is accountable to the
   A. fire chief.
   B. insurance companies.
   C. leaders of the governing body.
   D. civil servants.

Vocabulary

Define the following terms using the space provided.

1. Safety officer:

2. Paramedic:

3. Incident commander (IC):

4. Company officer:

5. Training officer:

Fundamentals of Fire Fighter Skills

Fill-in
Read each item carefully, and then complete the statement by filling in the missing word(s).

1. Each fire department is responsible for a specific _____ area.

2. When multiple agencies work together, a unified command system must be established. This system is referred to as the _____ _____ _____.

3. _____ _____ _____ provide specific information on the actions that should be taken to accomplish a certain task.

4. The first fire regulations in North America were established in Boston, Massachusetts, when the city banned _____ _____ and _____ _____.

5. _____ _____ _____ personnel administer prehospital care to people who are sick and injured.

6. The first volunteer fire company began in Philadelphia in 1735, under the leadership of _____ _____.

7. George Smith, a fire fighter in New York City, developed the first _____ _____ in 1817.

8. The _____ developed the first municipal water systems.

9. In Washington, D.C., _____ _____ _____ were introduced as the first communication tool used to send coded telegraph signals to the fire departments.

True/False
If you believe the statement to be more true than false, write the letter "T" in the space provided. If you believe the statement to be more false than true, write the letter "F."

_____ 1. Captains report directly to chiefs.

_____ 2. Covering a fire to ensure a low burn is called "banking."

_____ 3. George Washington established one of the first fire departments in Alexandria, Virginia, in 1765.

_____ 4. The fire fighter is responsible for dispatching units to an incident.

_____ 5. The Peshtigo fire storm jumped the 60-mile-wide Green Bay and continued to burn on Wisconsin's northeast peninsula.

_____ 6. The organizational structure of a fire department consists of a division of labor.

_____ 7. The battalion chief is the second rank of promotion, responsible for managing a fire company.

_____ 8. "Info techs" serve as liaisons between the IC and the news media.

_____ 9. Today almost all fire protection in the United States is funded directly or indirectly through tax dollars.

_____ 10. Consensus documents are developed through agreement among people representing different organizations and interests.

Short Answer

Complete this section with short written answers using the space provided.

1. Identify and describe the role of five companies common to most fire departments.

2. Describe the differences among regulations, policies, and standard operating procedures.

3. Identify the four basic management principles utilized in most fire departments.

4. Identify and describe 10 of the common and/or specialist positions a fire fighter may assume in his or her career as a fire fighter.

5. Outline the roles and responsibilities of a Fire Fighter II.

Fundamentals of Fire Fighter Skills

Word Fun

The following crossword puzzle is an activity provided to reinforce correct spelling and understanding of terminology associated with firefighting. Use the clues provided to complete the puzzle. Do not include spaces or punctuation when filling in the puzzle.

Clues

Across

1. An atmosphere-supplying respirator that supplies a respirable air atmosphere to the user from a breathing air source that is independent of the ambient environment and designed to be carried by the user. (NFPA 1981)
3. Covering a fire to ensure low burning.
7. The process by which an organization exercises authority and performs the functions assigned to it.
9. A member of EMS who can perform limited procedures that usually fall between those provided by an EMT and those provided by a Paramedic, including IV therapy, interpretation of cardiac rhythms, defibrillation, and airway intubation.
12. Historically, an identifying symbol on a building informing fire fighters that the building was insured by a company that would pay them for extinguishing the fire.
13. An obsolete amplification device that enabled a chief officer to give orders to fire fighters during an emergency. It was a precursor to a bullhorn and portable radios.
14. To take off an item of clothing or equipment.
15. Formal statements that provide guidelines for present and future actions.
16. The second rank of promotion in the fire service, between the lieutenant and the battalion chief. They are responsible for managing a fire company and for coordinating the activities of that company among the other shifts.
17. The guidelines that a department sets for fire fighters to work within.

Down

2. Usually the first level of fire chief; also called a district chief. These chiefs are often in charge of running calls and supervising multiple stations or districts within a city.
4. To put on an item of clothing or equipment.
5. A company officer who is usually responsible for a single fire company on a single shift; the first in line among company officers.
6. A tool used to pull down burning structures.
8. A rank structure, spanning the fire fighter through the fire chief, for managing a fire department and fire-ground operations.
10. EMS personnel with the highest level of training in EMS, including cardiac monitoring, administering drugs, inserting advanced airways, manual defibrillation, and other advanced assessment and treatment skills.
11. A valve installed to control water accessed from wooden pipes.
12. Members of the fire department who protect fire fighters by controlling traffic and securing the scene from public access.

Fire Alarms

The following real case scenarios will give you an opportunity to explore the concerns associated with the history and orientation of the fire service. Read each scenario, then answer each question in detail.

1. You have chosen the fire services as a career, and you have worked hard to get to this point. You have successfully completed the entry requirement, and you have been issued your bunker gear and uniform. You need to keep yourself on target to become a proud and accomplished fire fighter. What do you need to do to succeed?

   _____
   _____
   _____
   _____
   _____
   _____

2. You are outside the fire station washing the fire truck when you are approached by three children on bicycles. None of the children is wearing a helmet. They ask you if you will show them the truck and the station. How will you proceed?

   _____
   _____
   _____
   _____
   _____
   _____
   _____

Fire Fighter II in Action

The following scenario will give you an opportunity to apply your firefighting knowledge and your fire department SOGs to the new information you learned while studying this chapter. Research your department's SOGs and answer the assignment(s) in detail. Compare your answers with your classmates' and discuss similarities and obvious differences between your answers.

Your company officer has informed you that a group of 12-year-old scouts will be taking a tour of your fire station in the afternoon. He has selected you to explain what an engine company is, the tools and equipment carried, and the company's job functions at a fire.

1. Prepare an outline that will assist you in this portion of the tour. You will be given about 45 minutes with the scouts.

   _____
   _____
   _____
   _____
   _____
   _____
   _____
   _____

Fire Fighter Safety

Workbook Activities

The following activities have been designed to help you. Your instructor may require you to complete some or all of these activities as a regular part of your fire fighter training program. You are encouraged to complete any activity that your instructor does not assign as a way to enhance your learning in the classroom.

Chapter Review

The following exercises provide an opportunity to refresh your knowledge of this chapter.

Matching

Match each of the terms in the left column to the appropriate definition in the right column.

_____ 1. Portable radio

_____ 2. Critical incident stress debriefing (CISD)

_____ 3. Employee assistance program (EAP)

_____ 4. Rapid intervention crew

_____ 5. Freelancing

_____ 6. Occupational Safety and Health Administration (OSHA)

_____ 7. Personnel accountability system

_____ 8. Rehabilitation

A. The dangerous practice of acting independently of command instructions

B. A minimum of two fully-equipped personnel who are on-site, in a ready state, for immediate rescue of injured or trapped fire fighters

C. Readily identifies the locations and functions of all fire fighters at an incident

D. The federal agency that regulates worker safety and, in some cases, responder safety

E. Fire service programs that provide confidential help to fire fighters with personal issues

F. A portable communication device used by fire fighters

G. A process to provide periods of rest and recovery for emergency workers during an incident

H. Postincident meeting designed to assist rescue personnel in dealing with psychological trauma

Multiple Choice

Read each item carefully, and then select the best response.

_____ 1. A process to provide periods of rest and recovery for emergency workers during an incident is called
   A. recon.
   B. rehabilitation.
   C. RIC.
   D. relegate.

CHAPTER 2

_____ 2. Heat exhaustion, if recognized early, can be remedied by
   A. rehydration.
   B. cooling.
   C. rest.
   D. all of the above.

_____ 3. According to the NFPA, approximately what percentage of fire fighter deaths occurs during training?
   A. 10 percent
   B. 15 percent
   C. 20 percent
   D. 25 percent

_____ 4. What is the second-most-common cause of fire fighter deaths?
   A. Explosions
   B. Exposure to diseases
   C. Motor vehicle crashes
   D. Roof collapses

_____ 5. The system in which two fire fighters work as a team for safety purposes is referred to as the
   A. incident management system.
   B. buddy system.
   C. personnel accountability system.
   D. personal alert system.

_____ 6. What is the leading cause of death among fire fighters?
   A. Vehicle accidents
   B. Smoke inhalation
   C. Heart attack
   D. Cancer

_____ 7. The written rules and procedures that outline how to perform various functions and operations are the
   A. general operating guidelines.
   B. NIOSH Codes.
   C. incident management system (IMS).
   D. standard operating procedures (SOPs).

_____ 8. It is estimated that one vehicle collision involving an emergency vehicle occurs for every how many emergency responses?
   A. 500
   B. 1000
   C. 2000
   D. 3000

_____ 9. The most common fire fighter line-of-duty injuries are
   A. strains.
   B. sprains.
   C. soft tissue injuries.
   D. all of the above.

_____ 10. Most fire fighter injuries and deaths are the result of
   A. equipment failure.
   B. preventable situations.
   C. burns and explosions.
   D. hazardous materials.

_____ 11. The National Fallen Firefighters Foundation has developed a fire fighter safety initiative called
   A. Everyone Goes Home.
   B. NFA Safety.
   C. NFPA Safety.
   D. OSHA and Safety.

_____ 12. An ideal fire fighter exercise routine should include
   A. weight training.
   B. cardiovascular workouts.
   C. stretching.
   D. all of the above.

_____ 13. Which ability comes only through experience and with well-developed competency in basic fire fighter skills?
   A. Observation skills
   B. Incident command knowledge
   C. Situational awareness
   D. RIC skills

_____ 14. Your life and the lives of your crew depend on
   A. good physical fitness.
   B. firefighting ropes.
   C. diet.
   D. the partner system.

Vocabulary

Define the following terms using the space provided.

1. Personnel accountability system:

   _____
   _____
   _____

2. Standard operating procedures (SOPs):

   _____
   _____
   _____

3. Employee assistance program (EAP):

   _____
   _____
   _____

4. The 16 Firefighter Life Safety Initiatives:

   _____
   _____
   _____

5. Incident safety officer:
_____
_____
_____

6. Buddy system:
_____
_____
_____

Fill-in

Read each item carefully, and then complete the statement by filling in the missing word(s).

1. _____ collisions are a major cause of fire fighter fatalities.

2. The _____ _____ _____ _____ has developed programs with the goal of reducing line-of-duty deaths.

3. _____ _____ _____ provide confidential help with a wide range of problems that might affect performance.

4. The _____ officer has the authority to stop any activity that is judged to be unsafe.

5. _____ _____ _____ will prevent most vehicle collisions.

6. The National Fire Fighter _____-_____ _____ _____ provides a method for reporting situations that could have resulted in injuries or deaths.

7. All fire fighters—whether career or volunteer—should spend at least _____ _____ each day in physical fitness training.

8. _____ _____ is the leading cause of death both in the United States as a whole and among fire fighters in particular.

9. When fire fighters make independent decisions or do not follow command instructions, they are taking part in the dangerous practice of _____.

10. Emergency vehicle operators are subject to all _____ _____, unless a specific exemption is made.

Fundamentals of Fire Fighter Skills

True/False

If you believe the statement to be more true than false, write the letter "T" in the space provided. If you believe the statement to be more false than true, write the letter "F."

_____ 1. A prompt response is a higher priority than a safe response.

_____ 2. Most fire departments have employee assistance programs to provide counseling services to support fire fighters.

_____ 3. Freelancing is a good method of discovering new firefighting techniques.

_____ 4. Most fire fighter injuries and deaths are the result of preventable situations.

_____ 5. Every fire department must have a personnel accountability system.

_____ 6. Members of rapid intervention teams are the first fire fighters to enter a structure in an emergency operation.

_____ 7. Even with an emergency driving exemption, the operator can be found criminally or civilly liable if involved in a crash.

_____ 8. Emergency vehicle operators are subject to all traffic regulations unless a specific exemption is made.

_____ 9. On the fire ground, the company officer must always know where his or her teams are and what they are doing.

_____ 10. Fire fighters need not be aware of their surroundings when performing their assigned tasks at an emergency scene.

Short Answer

Complete this section with short written answers using the space provided.

1. Identify five of the nine Guidelines for Safe Emergency Vehicle Response.

   _____
   _____
   _____
   _____
   _____

2. Describe the purpose of a critical incident stress debriefing (CISD).

   _____
   _____
   _____
   _____
   _____

3. Identify four guidelines to stay safe, both on and off the job.

4. Identify the four major components of a successful safety program.

5. Identify three groups that fire fighters must always consider when ensuring safety at the scene.

Fundamentals of Fire Fighter Skills

Word Fun

The following crossword puzzle is an activity provided to reinforce correct spelling and understanding of terminology associated with firefighting. Use the clues provided to complete the puzzle. Do not include spaces or punctuation when filling in the puzzle.

Clues

Across

4. The _____ crew includes a minimum of two fully equipped personnel who are on-site, in a ready state, for immediate rescue of injured or trapped fire fighters.
6. An intervention designed to mitigate against the physical, physiological, and emotional stress of firefighting so as to sustain a member's energy, improve performance, and decrease the likelihood of on-scene injury or death. (NFPA 1521)

Down

1. The personnel _____ system readily identifies both the locations and the functions of all members operating at an incident scene. (NFPA 1500)
2. A postincident meeting designed to assist rescue personnel in dealing with psychological trauma as the result of an emergency. (NFPA 1006)
3. The dangerous practice of acting independently of command instructions.
5. _____ assistance programs provide confidential help to fire fighters with personal issues.
7. The federal agency that regulates worker safety and, in some cases, responder safety; part of the U.S. Department of Labor.

Fire Alarms

The following real case scenario will give you an opportunity to explore the concerns associated with fire fighter qualifications and safety. Read the scenario, and then answer the question in detail.

1. Your company officer has requested you give a morning training lecture on the second most common cause of fire fighter deaths. Identify this common cause of fire fighter death, and outline some of the information you can use to support your lecture. What can you do to reduce the chance of death or injury from this common cause of fire fighter death?

   _____
   _____
   _____
   _____
   _____
   _____

Fire Fighter II in Action

The following scenario will give you an opportunity to apply your firefighting knowledge and your fire department SOGs to the new information you learned while studying this chapter. Research your department's SOGs and answer the assignment(s) in detail. Compare your answers with your classmates' and discuss similarities and obvious differences between your answers.

1. Your company officer has directed you to make a presentation to your company on fire fighter deaths. Using resources available to your department, break down the statistics and deliver a program discussing fire fighter deaths by type of duty, cause of injury, and nature of injury.

   _____
   _____
   _____
   _____
   _____
   _____

Personal Protective Equipment and Self-Contained Breathing Apparatus

Workbook Activities

The following activities have been designed to help you. Your instructor may require you to complete some or all of these activities as a regular part of your fire fighter training program. You are encouraged to complete any activity that your instructor does not assign as a way to enhance your learning in the classroom.

Chapter Review

The following exercises provide an opportunity to refresh your knowledge of this chapter.

Matching

Match each of the terms in the left column to the appropriate definition in the right column.

_____ 1. Personal alert safety system
_____ 2. Cascade system
_____ 3. Self-contained breathing apparatus (SCBA)
_____ 4. Nomex®
_____ 5. IDLH
_____ 6. Don
_____ 7. Doff
_____ 8. Compressor
_____ 9. Open-circuit breathing apparatus
_____ 10. Face piece

A. A flame-resistant synthetic material
B. To put on an item of clothing or equipment
C. To take off an item of clothing or equipment
D. Several large storage cylinders of compressed breathing air connected by a high-pressure manifold system
E. Filters atmospheric air, compresses it to a high pressure, and transfers it to the SCBA cylinders
F. An electronic device that sounds a loud audible signal when a fire fighter becomes trapped or injured
G. An apparatus in which a tank of compressed air provides the breathing air supply for the user, and exhaled air is released into the atmosphere through a one-way valve
H. A component of the SCBA that covers the entire face
I. Provides respiratory protection by giving the fire fighter an independent, limited air supply
J. Any condition that would pose an immediate or delayed threat to life, cause irreversible adverse health effects, or interfere with an individual's ability to escape unaided from a hazardous environment

CHAPTER 3

Multiple Choice

Read each item carefully, and then select the best response.

_____ 1. An electronic semiconductor that emits a single-color light when activated is an LED, which is an acronym for
   A. light-emergent device.
   B. light-exiting device.
   C. light-emitting diode.
   D. laser-emitting diode.

_____ 2. PBI®, Kevlar®, and Nomex® are materials used in the construction of
   A. personal protective clothing.
   B. firefighting ropes.
   C. communications equipment.
   D. the buddy system.

_____ 3. In general, an SCBA weighs at least
   A. 25 pounds (11.34 kilograms).
   B. 30 pounds (13.61 kilograms).
   C. 40 pounds (18.14 kilograms).
   D. 45 pounds (20.41 kilograms).

_____ 4. An SCBA in which exhaled air is released into the atmosphere and is not reused is a(n)
   A. closed-circuit breathing apparatus.
   B. cascade system.
   C. open-circuit breathing apparatus.
   D. supplied-air apparatus.

_____ 5. What provides the frame for mounting the other working parts of the SCBA?
   A. Regulator
   B. Backpack
   C. Harness
   D. Straps and belt

_____ 6. During operating mode, if the SCBA regulator fails to function properly, what releases a constant flow of breathing air into the face piece?
   A. Respirator
   B. Air line
   C. Purge valve
   D. PBI®

_____ 7. Putting on an item of clothing or equipment is called
   A. doffing.
   B. freelancing.
   C. prepping.
   D. donning.

_____ 8. For structural firefighting, only use turnout coats that meet which NFPA standard?
   A. NFPA 1776
   B. NFPA 1492
   C. NFPA 1971
   D. NFPA 1963

Fundamentals of Fire Fighter Skills

_____ 9. Which of the following is designed to help colleagues locate a downed fire fighter by sending out a loud audible signal?
   A. Respirator
   B. PASS device
   C. SCBA regulator
   D. LED

_____ 10. Oxygen deficiency occurs when the atmosphere's oxygen level drops below
   A. 19.5 percent.
   B. 21 percent.
   C. 9 percent.
   D. 6 percent.

_____ 11. Donning protective clothing must be done in a specific order, and quickly. Fire fighters should be able to don protective clothing in
   A. 30 seconds.
   B. 45 seconds.
   C. 60 seconds.
   D. 120 seconds.

_____ 12. When plastic products burn, one of the most dangerous gases produced is
   A. hydrogen cyanide.
   B. phosgene gas.
   C. hydrochloric gas.
   D. phosphorous gas.

_____ 13. The part of the SCBA face piece that comes in contact with the skin is made of
   A. rubber or silicon.
   B. a cascade system.
   C. cotton.
   D. Nomex®.

_____ 14. The U.S. Department of Transportation requires this kind of testing on a periodic basis to ensure that SCBA cylinders are in good working condition.
   A. Thread
   B. Hydrostatic
   C. NIOSH
   D. Standard operating procedure

_____ 15. An SCBA designed to recycle the user's exhaled air is a(n)
   A. closed-circuit breathing apparatus.
   B. cascade system.
   C. open-circuit breathing apparatus.
   D. supplied-air apparatus.

_____ 16. The interior atmosphere of a burning building is considered to be
   A. safe.
   B. warm.
   C. NIMS.
   D. IDLH.

_____ 17. Which part of the personal protective equipment is worn over the head and under the helmet to protect the neck and ears?
   A. Helmet shell
   B. Protective hood
   C. Bunker hood
   D. Face piece

_____ 18. The straps and fasteners used to attach the SCBA to the fire fighter are part of the
   A. face piece.
   B. backpack.
   C. bunker coat.
   D. harness.

_____ 19. The device on an SCBA that measures and displays the amount of pressure currently in the cylinder is the
   A. personal safety gauge.
   B. pressure gauge.
   C. SCBA regulator.
   D. air line.

_____ 20. What very toxic gas is commonly present in smoke?
   A. Carbon monoxide
   B. Hydrogen cyanide
   C. Phosgene
   D. All of the above

Vocabulary

Define the following terms using the space provided.

1. Smoke particles:

2. Oxygen deficiency:

3. National Institute for Occupational Safety and Health (NIOSH):

4. Immediately dangerous to life and health (IDLH):

5. Supplied-air respirator (SAR):

6. End-of-service-time indicator (EOSTI):

7. Hydrostatic testing:

_____
_____
_____

8. Cascade system:

_____
_____
_____

Fill-in

Read each item carefully, and then complete the statement by filling in the missing word(s).

1. A(n) _____ _____ _____ is a breathing apparatus that uses an external source for the breathing air.

2. Due to the risk of fire at the scene of a vehicle extrication incident, members of the emergency team will wear _____ _____ _____.

3. A(n) _____-_____ breathing apparatus is typically used for structural firefighting.

4. Smoke contains _____, substances capable of causing cancer.

5. _____ protective clothing must be done in a specific order to obtain maximum protection.

6. Most fires result in _____ _____ and produce a variety of by-products.

7. A fire fighter should always carry a(n) _____ _____, given that most interior firefighting takes place in near-dark, zero visibility conditions.

8. An atmosphere is described as _____ _____ when the oxygen level is 19.5 percent or less.

9. A fire fighter's _____ _____ _____ must provide full-body coverage and protection from a variety of hazards.

10. Composite-fiber-wrapped SCBA cylinders must be replaced every _____ _____.

True/False

If you believe the statement to be more true than false, write the letter "T" in the space provided. If you believe the statement to be more false than true, write the letter "F."

_____ 1. Wearing PPE decreases normal sensory abilities.

_____ 2. A used SCBA air cylinder can never be replaced with a full cylinder in the field.

_____ 3. A complete annual inspection and maintenance procedure must be performed on each SCBA unit.

_____ 4. Carbon monoxide is a toxic gas produced through incomplete combustion.

_____ 5. Structural firefighting gear is not designed for extended wildland firefighting.

_____ 6. A structural firefighting ensemble protects against many of the hazards present at a vehicle extrication incident, such as broken glass and sharp metal objects.

_____ 7. Some of the toxic droplets in smoke can cause poisoning if they are absorbed through the skin.

_____ 8. Phosgene gas is a dangerous by-product of incomplete combustion.

_____ 9. At 21 percent oxygen concentration in air, judgment and coordination is impaired and a fire fighter will have a lack of muscle control.

_____ 10. SCBA regulators control the flow of air by increasing the low pressure in the cylinder to a usable higher pressure for the user.

Short Answer

Complete this section with short written answers using the space provided.

1. What are the six types of protection provided by PPE?

2. Identify three types of flame-resistant material commonly used in the construction of firefighting PPE.

3. List three reasons why fire fighters need respiratory protection during fire incidents.

4. List the physiological effects of reduced oxygen concentrations.

5. Identify five limitations of PPE.

CLUES

Across

1. A toxic gas produced by the combustion of materials containing cyanide.
5. A device used for increasing the pressure and density of a gas. (NFPA 853)
8. The U.S. federal agency that regulates worker safety and, in some cases, responder safety. It is part of the U.S. Department of Labor.
9. A component of SCBA that fits over the face.
10. A respirator with an independent air supply that is used by underwater divers.
12. To take off an item of clothing or equipment.
13. A device that provides respiratory protection for the wearer. (NFPA 1994)
15. A warning device on a SCBA that alerts the user that the end of the breathing air is approaching.
17. The protective trousers worn by a fire fighter for interior structural firefighting.

Down

2. A method of piping air tanks together to allow air to be supplied to the SCBA fill station using a progressive selection of tanks, each with a higher pressure level. (NFPA 1901)
3. The component of the SCBA that stores the compressed air supply.
4. Any condition that would pose an immediate or delayed threat to life, cause irreversible adverse health effects, or interfere with an individual's ability to escape unaided from a hazardous environment. (NFPA 1670, 2004)
6. The U.S. federal agency responsible for research and development on occupational safety and health issues.
7. A chemical agent that causes severe pulmonary damage; it is a by-product of incomplete combustion.
11. The hose through which air flows, either within an SCBA or from an outside source to a supplied air respirator.
14. A device that continually monitors for lack of movement of the wearer and, if no movement is detected, automatically activates an alarm signal indicating the wearer is in need of assistance. The device can also be manually activated to trigger the alarm signal. (NFPA 1982)
16. To put on an item of clothing or equipment.

Word Fun

The following crossword puzzle is an activity provided to reinforce correct spelling and understanding of terminology associated with firefighting. Use the clues provided to complete the puzzle. Do not include spaces or punctuation when filling in the puzzle.

Fire Alarms

The following real case scenarios will give you an opportunity to explore the concerns associated with fire fighter qualifications and safety. Read each scenario, and then answer each question in detail.

1. It is just after dinner when your ladder truck company is dispatched to an apartment fire. Upon arrival, you are assigned to search and rescue on the second floor. You have just completed searching the first apartment when your SCBA regulator malfunctions. How should you proceed?

2. You have just returned from a commercial structure fire at a mattress store. Your PPE is soiled with the products of combustion. Although you were sprayed off with water at the scene, your gear is still very dirty. How will you proceed?

Fundamentals of Fire Fighter Skills

Fire Fighter II in Action

The following scenario will give you an opportunity to apply your firefighting knowledge and your fire department SOGs to the new information you learned while studying this chapter. Research your department's SOGs and answer the assignment(s) in detail. Compare your answers with your classmates' and discuss similarities and obvious differences between your answers.

1. Explain why it is important that fire fighters keep their turnout gear and other PPE clean and in good working order.

   _____
   _____
   _____
   _____
   _____
   _____

2. What are your department SOGs for PPE cleaning and maintenance?

   _____
   _____
   _____
   _____
   _____
   _____

Skill Drills

Skill Drill 3-1: Donning Personal Protective Clothing

Test your knowledge of this skill drill by placing the photos below in the correct order. Number the first step with a "1," the second step with a "2," and so on.

© Jones & Bartlett Learning. Photographed by Glen E. Ellman.

_____ Place your equipment in a logical order for donning.

© Jones & Bartlett Learning. Photographed by Glen E. Ellman.

_____ Have your partner check your clothing.

© Jones & Bartlett Learning. Photographed by Glen E. Ellman.

_____ Put on your boots and pull up your bunker pants. Place the suspenders over your shoulders and secure the front of the pants.

© Jones & Bartlett Learning. Photographed by Glen E. Ellman.

_____ Place your protective hood over your head and down around your neck.

© Jones & Bartlett Learning. Photographed by Glen E. Ellman.

_____ Place your helmet on your head and adjust the chin strap securely. Turn up your coat collar and secure it in front.

© Jones & Bartlett Learning. Photographed by Glen E. Ellman.

_____ Put on your gloves.

© Jones & Bartlett Learning. Photographed by Glen E. Ellman.

_____ Put on your turnout coat and close the front of the coat.

Fire Service Communications

Workbook Activities

The following activities have been designed to help you. Your instructor may require you to complete some or all of these activities as a regular part of your fire fighter training program. You are encouraged to complete any activity that your instructor does not assign as a way to enhance your learning in the classroom.

Chapter Review

The following exercises provide an opportunity to refresh your knowledge of this chapter.

Matching

Match each of the terms in the left column to the appropriate definition in the right column.

_____ 1. Time marks
_____ 2. CAD
_____ 3. Mobile data terminals
_____ 4. Portable radio
_____ 5. Radio repeater system
_____ 6. ANI
_____ 7. Duplex channel
_____ 8. Talk-around channel
_____ 9. Simplex channel
_____ 10. Mayday

A. A battery-operated, hand-held transceiver
B. Update that should include the type of operation, the progress of the incident, the anticipated actions, and the need for additional resources
C. Allow fire fighters to receive information in the apparatus or at the station
D. Code indicating that a fire fighter is lost, is missing, or requires immediate assistance
E. A radio channel using two frequencies
F. Automated systems used by telecommunicators to obtain and assess dispatch information
G. A radio system that automatically retransmits a radio signal on a different frequency
H. A radio channel using one frequency
I. A radio channel that bypasses a repeater system
J. Automatic number identification

Multiple Choice

Read each item carefully, and then select the best response.

_____ 1. The process of assigning a response category is based on the nature of the reported problem or
   A. classification and prioritization.
   B. location validation.
   C. unit selection.
   D. dispatch directive.

CHAPTER 4

_____ 2. Some dispatch centers are equipped with automatic vehicle locator systems that track apparatus by using
   A. GPS devices.
   B. CAD systems.
   C. FCC regulated technology.
   D. TDD systems.

_____ 3. Two-way radios that are permanently mounted in vehicles are called
   A. portable radios.
   B. base stations.
   C. simplex channel radios.
   D. mobile radios.

_____ 4. Telecommunicators must follow standard operating procedures (SOPs) and use
   A. the incident management system.
   B. active listening.
   C. FCC guidelines.
   D. talk-around channels.

_____ 5. One of the first things you should learn when assigned to a fire station is how to use the
   A. personal protective equipment (PPE).
   B. response vehicles.
   C. incident management system.
   D. telephone and intercom system.

_____ 6. The central processing point for all information relating to an emergency incident is the
   A. incident commander.
   B. communications center.
   C. fire department.
   D. computer-aided dispatch.

_____ 7. The first-arriving unit at an incident should always give a brief initial report and
   A. control traffic.
   B. determine the duration of the ongoing incident.
   C. establish command.
   D. prepare an offensive attack unit.

_____ 8. Unit selection is the process of determining exactly which
   A. radio frequency to assign.
   B. equipment will be needed in the response.
   C. attack strategy will be assigned.
   D. unit(s) to dispatch.

_____ 9. A CAD system helps meet the most important objective in processing an emergency call, which is
   A. recording communications messages.
   B. documenting the incident.
   C. sending the appropriate units to the correct location as quickly as possible.
   D. identifying the potential casualties.

_____ 10. The telecommunicator can initiate a response after determining the
   A. location and nature of the problem.
   B. time of the communication and nature of the problem.
   C. urgency of the response.
   D. fire department location.

_____ 11. Someone in the communications center must remain in contact with the responding units
   A. until an incident commander is on the scene.
   B. throughout the incident.
   C. throughout the on-site scene assessment.
   D. until the incident is under control.

_____ 12. A call box connects a person directly to a(n)
   A. fire department.
   B. police station.
   C. incident commander.
   D. telecommunicator.

_____ 13. The first-arriving unit at an incident should
   A. give a brief initial radio report.
   B. establish command.
   C. tell other responding units what is happening.
   D. all of the above.

_____ 14. Radio codes, such as "ten codes,"
   A. are widely used and popular.
   B. are understood by all radio operators.
   C. can be problematic.
   D. work well when responding with other jurisdictions.

_____ 15. When you speak into the microphone, always speak across the microphone
   A. at a 90-degree angle.
   B. at a 45-degree angle, holding the microphone 1 to 2 inches (2.5 to 5 cm) from the mouth.
   C. as loudly as possible.
   D. without background interference.

_____ 16. A group of shared frequencies controlled by a computer is called a
   A. trunking system.
   B. mobile radio system.
   C. radio repeater system.
   D. base station system.

_____ 17. Call classification determines the
   A. incident management system.
   B. equipment to transport to the incident.
   C. record documentation format.
   D. number and types of units that are dispatched.

_____ 18. Urgent messages that take priority over all other communications are known as
   A. time marks.
   B. dispatch information.
   C. emergency traffic.
   D. ten-code communications.

_____ 19. Which of the following is used to warn all fire personnel to pull back to a safe location?
   A. Evacuation signal
   B. Mayday
   C. Retreat signal
   D. PAR

Vocabulary

Define the following terms using the space provided.

1. Automatic location identification:

2. Run cards:

3. TDD/TTY/text phone:

4. Ten-codes:

5. Time marks:

6. Activity logging system:

7. Computer-aided dispatch (CAD):

8. Evacuation signal:

Fill-in
Read each item carefully, and then complete the statement by filling in the missing word(s).

1. A(n) _____ sends out emergency response resources promptly to an address or incident location for a specific purpose.

2. Radio systems that use one frequency to transmit and receive all messages are called _____ channels.

3. A(n) _____ channel radio utilizes two frequencies per channel.

4. A(n) _____ radio is a two-way radio that is permanently mounted in a fire apparatus.

5. A(n) _____ system uses a shared bank of frequencies to make the most efficient use of radio resources.

6. The facility that receives the emergency reports and is responsible for dispatching fire department units is the _____ _____.

7. The _____ is a trained individual responsible for answering requests for emergency assistance from citizens.

8. _____ _____ provide status updates to the communications center at predetermined intervals.

9. _____ is an emergency code indicating that a fire fighter is missing or requires immediate assistance.

10. A(n) _____ signal warns all personnel to pull back to a safe location.

True/False
If you believe the statement to be more true than false, write the letter "T" in the space provided. If you believe the statement to be more false than true, write the letter "F."

_____ 1. Size-up should be transmitted by the first-arriving unit at an incident.

_____ 2. Individuals with speech or hearing impairments can access the 911 system through telephones.

_____ 3. All calls to 911 are directed to a designated public safety answering point for that jurisdiction.

_____ 4. The telecommunicator who takes a call must conduct a "telephone interrogation."

_____ 5. A telecommunicator can initiate a response with just two pieces of information—the location and nature of the problem.

_____ 6. Before transmitting over a radio, push and hold the PTT button for at least 2 seconds.

_____ 7. Most fire departments use plain English for radio communications.

_____ 8. Automatic number identification displays the telephone number where the call originated.

_____ 9. The police, fire, and EMS departments must always have separate communication centers.

_____ 10. Municipal fire alarm boxes are the most reliable source of contact with the communications center to indicate an emergency.

Short Answer

Complete this section with short written answers using the space provided.

1. What are five of the basic functions performed in a communications center?

2. List five pieces of equipment often found in a communications center.

3. Identify the five major steps in processing an emergency incident.

4. Why is accuracy important in report documentation?

5. Define *emergency traffic*.

6. Describe common evacuation signals.

7. Explain how to initiate a mayday call.

Fundamentals of Fire Fighter Skills

CLUES

Across

1. A telephone that connects two predetermined points.
5. A system of telephones connected by phone lines, radio equipment, or cellular technology to a communications center or fire department.
6. A stationary radio transceiver with an integral AC power supply. (NFPA 1221)
9. A radio system that uses two frequencies per channel: one frequency transmits and the other receives messages.
11. Cards used to determine a predetermined response to an emergency.
12. A radio system that uses a shared bank of frequencies to make the most efficient use of radio resources.

Down

1. To send out emergency response resources promptly to an address or incident location for a specific purpose. (NFPA 450)
2. A system of predetermined coded messages, such as "What is your 10–20?", used by responders over the radio.
3. A two-way radio that is permanently mounted in a fire apparatus.
4. A code indicating that a fire fighter is lost, missing, or trapped, and requires immediate assistance.
7. Status updates provided to the communications center every 10 to 20 minutes. Such an update should include the type of operation, the progress of the incident, the anticipated actions, and the need for additional resources.
8. A system used by fire departments to report and maintain computerized records of fires and other fire department incidents in a uniform manner.
10. A facility in which 911 calls are answered. (NFPA 1221)

Word Fun

The following crossword puzzle is an activity provided to reinforce correct spelling and understanding of terminology associated with firefighting. Use the clues provided to complete the puzzle. Do not include spaces or punctuation when filling in the puzzle.

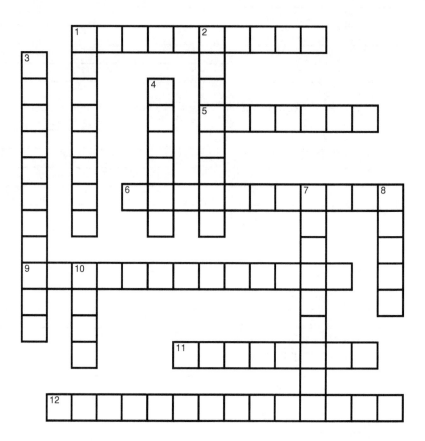

Fire Alarms

The following real case scenarios will give you an opportunity to explore the concerns associated with fire service communications. Read each scenario, and then answer each question in detail.

1. You are sitting in the report room finishing up paperwork from an earlier call when the phone rings. You answer the phone using your department's standard operating procedures (SOPs). The caller states that her neighbor's burn pile has grown out of control and that it is rapidly getting larger. The caller states that her neighbor is chasing the fire with a garden hose. How should you proceed?

   _____
   _____
   _____
   _____
   _____
   _____

2. Your engine company is dispatched to a commercial structure fire in a large grocery store. This is the second fire to which you have responded on your shift. You and your crew are assigned to a search and rescue team in the storage area of the store. You become disoriented and cannot find your way out. You have your radio and need to contact the IC to let him know you are lost. How will you proceed?

   _____
   _____
   _____
   _____
   _____
   _____

Fire Fighter II in Action

The following scenario will give you an opportunity to apply your firefighting knowledge and your fire department SOGs to the new information you learned while studying this chapter. Research your department's SOGs and answer the assignment in detail. Compare your answers with your classmates' and discuss similarities and obvious differences between your answers.

Your team is on the third floor of a three-story residential structure performing a search for a missing occupant. You have a charged hoseline in the stairwell, and an axe and halligan. You have completed one-half of a left hand search pattern when the IC calls for an emergency evacuation.

1. What are your priorities at this time?

2. How will you safely and quickly evacuate this structure?

Skill Drills

Skill Drill 4-1: Initiating a Response to a Simulated Emergency
Test your knowledge of the skill drill by filling in the correct words in the photo captions.

© Jones & Bartlett Learning. Photographed by Glen E. Ellman.

1. Identify your agency. Ask whether there is a(n) _____. Organize your questions to get the following information: incident _____, type of incident, and when the incident occurred. Obtain the following information: caller's _____; location of the caller, if different from the incident; and _____ number.

© Jones & Bartlett Learning. Photographed by Glen E. Ellman.

2. Record the information needed. Initiate a(n) _____ following the protocols of your _____ _____.

Incident Command System

Workbook Activities

The following activities have been designed to help you. Your instructor may require you to complete some or all of these activities as a regular part of your fire fighter training program. You are encouraged to complete any activity that your instructor does not assign as a way to enhance your learning in the classroom.

Chapter Review

The following exercises provide an opportunity to refresh your knowledge of this chapter. All questions in this chapter are Fire Fighter II level.

Matching

Match each of the terms in the left column to the appropriate definition in the right column.

_____ 1. Staging area   A. Usually refers to companies and/or crews working in the same geographic area

_____ 2. Task force   B. A supervisory level established to manage the span of control above the division, or group, level

_____ 3. Safety officer   C. A location close to the incident scene where a number of units can be held in reserve, ready to be assigned if needed

_____ 4. Branch   D. An organized group of fire fighters, working without an apparatus, under the leadership of a company officer

_____ 5. Transfer of command   E. The officer in charge of a fire department company

_____ 6. Single command   F. Includes two to five single resources, such as different types of units assembled to accomplish a specific task

_____ 7. Division   G. Usually refers to companies and/or crews working on the same task or function

_____ 8. Crew   H. Position responsible for identifying and evaluating hazards or unsafe conditions at an incident

_____ 9. Company officer   I. The most traditional perception of the Command function; the genesis of the term *incident commander*.

_____ 10. Group   J. Occurs when one person relinquishes command to another individual

CHAPTER 5

Multiple Choice

Read each item carefully, and then select the best response.

_____ 1. The incident commander's (IC's) representative and the position responsible for exchanging information with outside agencies or directing people to the proper authority is the
   A. liaison officer.
   B. safety officer.
   C. public information officer.
   D. Planning Section Chief.

_____ 2. The exterior sides of a building are generally known as sides
   A. north, east, south, and west.
   B. A, B, C, and D.
   C. one, two, three, and four.
   D. right, left, top, and bottom.

_____ 3. The only position in the incident command system (ICS) that must always be filled is
   A. safety officer.
   B. Command.
   C. liaison officer.
   D. public information officer.

_____ 4. A standard system of assigning and keeping track of the resources involved in the incident is the
   A. incident resources system.
   B. resource management system.
   C. unified resources system.
   D. operational resources system.

_____ 5. An ICS developed in the 1970s for day-to-day fire department incidents was the
   A. Unified Command System.
   B. FIRESCOPE.
   C. Fire-ground Command System.
   D. Task Force System.

_____ 6. When one person relinquishes command of an incident and another individual becomes the IC, there is a(n)
   A. transfer of command.
   B. passing command.
   C. operational transfer.
   D. resource transfer.

_____ 7. The section of the ICS that is responsible for providing supplies, services, facilities, and materials during the incident is the
   A. Operations Section.
   B. Planning Section.
   C. Logistics Section.
   D. Administration Section.

_____ 8. Which position is established when the first-arriving unit arrives on the scene?
   A. Command
   B. Division supervisor
   C. Operations Section Chief
   D. ICS director

_____ 9. The number of subordinates who report to one supervisor at any level within the organization is the
   A. integrated communications.
   B. unified command.
   C. unity of command.
   D. span of control.

_____ 10. Five units of the same type with an assigned leader is referred to as a(n)
   A. operating unit.
   B. strike team.
   C. branch.
   D. sector.

_____ 11. The section of the ICS that is responsible for the collection, evaluation, dissemination, and use of information relevant to the incident is the
   A. Operations Section.
   B. Planning Section.
   C. Logistics Section.
   D. Administration Section.

_____ 12. What is used to ensure that everyone at an emergency site can transfer information among both their supervisors and subordinates?
   A. Integrated communications
   B. Span of control
   C. Universal communications
   D. Command staff

_____ 13. The section of the ICS that is responsible for the management of all actions that are directly related to controlling the incident is the
   A. Operations Section.
   B. Planning Section.
   C. Logistics Section.
   D. Administration Section.

_____ 14. The section of the ICS that is responsible for any legal issues that may arise during an incident is the
   A. Operations Section.
   B. Planning Section.
   C. Logistics Section.
   D. Finance/Administration Section.

_____ 15. What is the management concept in which each person has only one direct supervisor?
   A. Integrated communications
   B. Unified command
   C. Unity of command
   D. Span of control

_____ 16. The IC, all direct support staff, and command functions should always be located at the
   A. management post.
   B. incident command post.
   C. Planning Section.
   D. staging area.

_____ 17. When multiple agencies with overlapping jurisdictions and legal responsibilities respond to the same incident, ICS may employ a(n)
   A. incident command system.
   B. unity of command.
   C. fire-ground command system.
   D. unified command.

_____ 18. Companies or crews working in the same geographic area are termed
   A. divisions.
   B. groups.
   C. sectors.
   D. teams.

_____ 19. The safety officer, liaison officer, and public information officer are always part of the
   A. command staff.
   B. rapid intervention team.
   C. Operations Section.
   D. ICS general staff.

_____ 20. A location close to the incident scene where a number of units can be held in reserve, ready to be assigned if needed, is the
   A. resources area.
   B. operations area.
   C. staging area.
   D. rapid intervention area.

_____ 21. Which officer has the authority to stop or suspend operations when unsafe situations occur?
   A. Health officer
   B. Liaison officer
   C. Operations officer
   D. Safety officer

_____ 22. The officer responsible for gathering and releasing incident information to the news media is the
   A. liaison officer.
   B. public information officer.
   C. Operations Section Chief.
   D. Public Relations Chief.

_____ 23. A crew is a group of fire fighters who are working
   A. on their own.
   B. as ICS Section Chiefs.
   C. without apparatus.
   D. outside the ICS.

_____ 24. An individual vehicle and its assigned personnel are considered a
   A. single resource.
   B. crew.
   C. branch.
   D. division.

Vocabulary

Define the following terms using the space provided.

1. Incident command system:

   _____
   _____
   _____

2. Unified command:

   _____
   _____
   _____

3. Incident action plan:

4. National Incident Management System:

5. Command staff:

6. Division:

7. Resource management:

Fill-in
Read each item carefully, and then complete the statement by filling in the missing word(s).

1. In the ICS structure, _____ is ultimately responsible for managing an incident and has the authority to direct all activities.

2. An ICS provides a standard _____, _____, and _____ procedure to organize and manage any operation.

3. A(n) _____ _____ is an individual vehicle and its assigned personnel.

4. The primary reason for establishing divisions and groups is to maintain a(n) effective _____ _____ _____.

5. Planning, supervision, and _____ are key components of an ICS.

6. Standard _____ is a strength of the ICS. Understanding this is the first step in understanding the system.

7. The modular design of ICS allows the _____ to expand based on the needs of the incident.

8. The staging, rehabilitation, and treatment areas are _____ areas where particular functions take place.

9. The staff of the four sections within the ICS are known as the ICS _____ _____.

10. Command is the only position in the ICS that must _____ be filled.

True/False

If you believe the statement to be more true than false, write the letter "T" in the space provided. If you believe the statement to be more false than true, write the letter "F."

_____ 1. The Logistics Section Chief reports directly to the Planning Section Chief.
_____ 2. Each block on an ICS organizational chart refers to an individual.
_____ 3. An engine and its crew are considered to be a single resource.
_____ 4. Prior to the ICS, the organization established to direct operations often varied based on the chief on duty.
_____ 5. The ICS is designed to be flexible and modular.
_____ 6. The Operations Section is responsible for updating the incident action plan.
_____ 7. An ICS provides a standard, professional, organized approach to managing emergency incidents.
_____ 8. The issues addressed by the Finance/Administration Section are usually addressed after the incident.
_____ 9. The incident action plan outlines the overall strategy for an emergency incident.
_____ 10. The different components of the ICS have different goals and objectives.

Short Answer

Complete this section with short written answers using the space provided.

1. Identify the four major functional components within the ICS.

   _____
   _____
   _____
   _____

2. When an officer relinquishes command, he or she needs to give the new IC an initial status report that includes which five pieces of information?

   _____
   _____
   _____
   _____
   _____

3. Identify 11 important characteristics of an ICS.

   _____
   _____
   _____
   _____

Fundamentals of Fire Fighter Skills

4. For which nine tasks is Command directly responsible?

5. List the five major functional components of an ICS.

Chapter 5: Incident Command System

Word Fun

The following crossword puzzle is an activity provided to reinforce correct spelling and understanding of terminology associated with firefighting. Use the clues provided to complete the puzzle. Do not include spaces or punctuation when filling in the puzzle.

CLUES

Across

2 The organizational level having functional, geographical, or jurisdictional responsibility for major aspects of incident operations. (NFPA 1026)
4 A supervisory level established to divide an incident into functional areas of operation. (NFPA 1561)
5 The incident _____ is a combination of facilities, equipment, personnel, procedures, and communications operating within a common organizational structure that has responsibility for the management of assigned resources to effectively accomplish stated objectives pertaining to an incident or training exercise. (NFPA 1670)
9 A team of two or more fire fighters. (NFPA 1500)
10 A prearranged, strategically placed area, where support response personnel, vehicles, and other equipment can be held in an organized state of readiness for use during an emergency. (NFPA 424)
12 A supervisory level established to divide an incident into geographic areas of operations. (NFPA 1561)
13 An individual, a piece of equipment and its personnel, or a crew or team of individuals with an identified supervisor that can be used on an incident or planned event. (NFPA 1026)

Down

1 The incident _____ is the field location at which the primary tactical-level, on-scene incident command functions are performed. (NFPA 1026)
3 The public information officer, safety officer, and liaison officer, all of whom report directly to the incident commander and are responsible for functions in the incident management system that are not a part of the function of the line organization. (NFPA 1561)
5 The first component of the ICS. It is the only position in the ICS that must always be staffed.
6 The section responsible for all tactical operations at the incident or planned event, including up to 5 branches; 25 divisions/groups; and 125 single resources, task forces, or strike teams. (NFPA 1026)
7 An organization of agencies established in the early 1970s to develop a standardized system for managing fire resources at large-scale incidents, such as wildland fires.
8 The section responsible for the collection, evaluation, dissemination, and use of information related to the incident situation, resource status, and incident forecast. (NFPA 1026)
11 A system mandated by Homeland Security Presidential Directive 5 (HSPD-5) that provides a systematic, proactive approach guiding government agencies at all levels, the private sector, and nongovernmental organizations to work seamlessly to prepare for, prevent, respond to, recover from, and mitigate the effects of incidents, regardless of cause, size, location, or complexity, so as to reduce the loss of life or property and harm to the environment. (NFPA 1026)

Fire Alarms

The following real case scenarios will give you an opportunity to explore the concerns associated with ICS. Read each scenario, and then answer each question in detail.

1. You are the team leader for an exposure crew on a fire in an apartment complex under construction. You are assigned to Division C, and your task is to protect an adjacent apartment building labeled exposure C. The Division D supervisor directs your crew to move your line and protect exposure D. How should you proceed?

   _____
   _____
   _____
   _____
   _____
   _____

2. It is a quiet Sunday afternoon when you are dispatched to an old four-story furniture warehouse located in the historical district of your city. Dispatch stated the fire started on the second floor. Upon arrival, you find fire pouring out of the A/D corner of the building, lapping into the third-floor windows. Your Lieutenant states that the preincident plan calls for the first engine to charge the standpipe. The Battalion Chief arrives on the scene, and your Lieutenant prepares to transfer command. How will she proceed?

   _____
   _____
   _____
   _____
   _____
   _____

Fire Fighter II in Action

The following scenario will give you an opportunity to apply your firefighting knowledge and your fire department SOGs to the new information you learned while studying this chapter. Research your department's SOGs and answer the assignment in detail. Compare your answers with your classmates' and discuss similarities and obvious differences between your answers.

Your company is the first to arrive on a working structure fire. As senior fire fighter, you take command, perform your 360, and give a size up. You direct your company to establish a water supply and prepare to make an offensive attack. The second dispatched company arrives within 5 minutes, and has a Captain on board.

1. How and why does a transfer of command take place?

2. As you relinquish command, what information should you include in your situation status report?

3. List the people the new IC should inform of this transfer of command.

Fire Behavior

Workbook Activities

The following activities have been designed to help you. Your instructor may require you to complete some or all of these activities as a regular part of your fire fighter training program. You are encouraged to complete any activity that your instructor does not assign as a way to enhance your learning in the classroom.

Chapter Review

The following exercises provide an opportunity to refresh your knowledge of this chapter.

Matching

Match each of the terms in the left column to the appropriate definition in the right column.

_____ 1. Endothermic
_____ 2. Gas
_____ 3. Decay
_____ 4. Oxidation
_____ 5. Conduction
_____ 6. Radiation
_____ 7. Plume
_____ 8. Convection
_____ 9. Nuclear fission
_____ 10. Hypoxia
_____ 11. Flash point
_____ 12. Heat flux
_____ 13. Incipient stage
_____ 14. Smoke explosion
_____ 15. Ventilation limited

A. The process of transferring heat through matter by movement of the kinetic energy from one particle to another
B. The column of hot gases, flames, and smoke rising above a fire; also called convection column, thermal updraft, or thermal column
C. One of the three states of matter
D. Transfer of heat through the emission of energy in the form of invisible waves
E. The process in which oxygen combines chemically with another substance to create a new compound
F. The stage of fire where the fire is running out of fuel or oxygen
G. Reactions that absorb heat or require heat to be added
H. Heat transfer by circulation within a medium such as a gas or a liquid
I. A state of inadequate oxygenation of the blood and tissue
J. Created by splitting the nucleus of an atom
K. A fire in an enclosed building that is restricted because there is insufficient oxygen available for the fire to burn as rapidly as it would with an unlimited supply of oxygen.
L. The measure of the rate of heat transfer from one surface to another.
M. A violent release of confined energy that occurs when a mixture of flammable gases and oxygen are present, usually in a void or other area separate from the fire compartment and come in contact with a source of ignition.
N. The lowest temperature at which a liquid produces a flammable vapor
O. The stage of fire development where the fire has not progressed beyond a size that can be extinguished with a portable fire extinguisher.

CHAPTER 6

Multiple Choice

Read each item carefully, and then select the best response.

_____ 1. A thin piece of wood burns quickly due to its
   A. mass.
   B. composition.
   C. weight-to-mass ratio.
   D. large surface area.

_____ 2. Which class of fires involves ordinary combustibles such as wood?
   A. Class A fires
   B. Class B fires
   C. Class C fires
   D. Class D fires

_____ 3. Matter is classified as:
   A. electricity, gasoline, natural gas
   B. Type A, Type B, Type C
   C. solid, liquid, gas
   D. Class A, Class B, Class C

_____ 4. A very rapid chemical process that combines oxygen with another substance and results in the release of heat and light is called
   A. oxidization.
   B. combustion.
   C. pyrolysis.
   D. decomposition.

_____ 5. Which class of fires involves flammable or combustible liquids such as gasoline?
   A. Class A fires
   B. Class B fires
   C. Class C fires
   D. Class D fires

_____ 6. The initial growth of a fire is largely dependent on:
   A. the type of fuel
   B. the amount of fuel being pyrolyzed into vapor
   C. thermal layering
   D. A and B

_____ 7. The movement of heat through a fluid medium such as air or a liquid is
   A. convection.
   B. endothermic.
   C. exothermic.
   D. conduction.

_____ 8. The phenomenon of gases forming into layers according to temperature is called
   A. thermal differentiation.
   B. thermal division.
   C. thermal layering.
   D. thermal balance.

_____ 9. The lowest temperature at which a liquid produces a flammable vapor is the
   A. flame point.
   B. fire point.
   C. ignition temperature.
   D. flash point.

_____ 10. In which stage of the fire does hot smoke and gases start to rise because of heating and becoming lighter?
   A. ignition stage
   B. growth stage
   C. fully developed stage
   D. decay stage

_____ 11. Which class of fires involves combustible metals?
   A. Class A fires
   B. Class B fires
   C. Class C fires
   D. Class D fires

_____ 12. As a fire grows, increasing volumes of hot gases and smoke create a
   A. tornado
   B. channel
   C. plume
   D. cloud

_____ 13. The decomposition of a material brought about by heat in the absence of oxygen is called
   A. evaporation.
   B. dehydration.
   C. decomposition.
   D. pyrolysis.

_____ 14. Incomplete combustion produces
   A. pure air.
   B. solids.
   C. smoke.
   D. oxidizers.

_____ 15. The weight of a gaseous fuel is the
   A. gas mass.
   B. vapor density.
   C. explosive limit.
   D. BLEVE.

_____ 16. When two materials rub together and produce friction, they create
   A. light energy.
   B. chemical energy.
   C. mechanical energy.
   D. potential energy.

_____ 17. The transfer of heat energy in the form of invisible waves is called
   A. radiation.
   B. oxidization.
   C. volatility.
   D. transpiration.

_____ 18. The four conditions that must be present for fire to take place are represented in the
   A. fire tetrahedron.
   B. fire square.
   C. fire triangle.
   D. fire rectangle.

_____ 19. Particles, vapors, and gases are the three major components of
   A. fumes.
   B. silt.
   C. smoke.
   D. exhaust.

_____ 20. The key to preventing a BLEVE is to
   A. flush the spill.
   B. ventilate the area.
   C. cool the top of the tank.
   D. apply an oxidizing agent.
_____ 21. A fire requires fuel that is in the form of
   A. combustible vapors
   B. solid
   C. liquid
   D. particles
_____ 22. The range of gas-air mixtures that will burn varies
   A. from one fuel to another
   B. with the amount of energy present
   C. with the vapor pressure
   D. with the vapor density

Labeling

Label the following diagram with the correct terms.

1. The fire tetrahedron

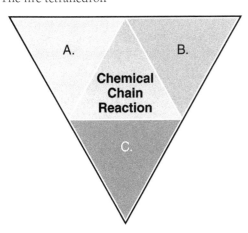

A. _____
B. _____
C. _____

Figure 6-4

Vocabulary

Define the following terms using the space provided.

1. Lower flammable limit:

2. Ignition temperature:

3. Flash point:

4. BLEVE:

5. Thermal layering:

6. Fire triangle:

7. Flashover:

8. Fully developed stage:

9. Rollover:

10. Smoke explosion:

Fill-in
Read each item carefully, and then complete the statement by filling in the missing word(s).

1. A fire involving a liquid fuel can be extinguished by shutting off the _____ of fuel, or using _____ to exclude oxygen from the fuel.

2. Carbon _____ is deadly in small quantities.

3. Research indicates that fires in modern residential occupancies are likely to enter a _____ _____ decay stage prior to the arrival of the first due engine company.

4. For a fuel to burn, it must be changed into a _____ _____

5. Because smoke is the product of incomplete combustion and contains unburned hydrocarbons, we need to remember that it is a form of _____.

6. When hot fire gases are exhausted from a fire building, if they are above the _____ _____ of the gases they may ignite upon mixing with a fresh supply of oxygen.

7. Matter exists in three states: _____, _____, and _____.

8. The _____ _____ is the level in a compartment opening at which the pressure of the hot gases and smoke leaving the compartment and the pressure of the cooler gases entering the compartment are equal.

9. The amount of liquid that is vaporized when it is heated relates to the _____ of the liquid.

10. Reactions that produce heat are referred to as _____ reactions.

True/False
If you believe the statement to be more true than false, write the letter "T" in the space provided. If you believe the statement to be more false than true, write the letter "F."

_____ 1. The size and shape of the fuel will greatly impact the ability of the fuel to ignite.

_____ 2. A fire can progress rapidly from the incipient stage to a growth stage in a very short period of time.

_____ 3. Flashover is a slow change or transition from the growth stage to the fully developed stage.

_____ 4. A column of hot black smoke coming into contact with an adequate supply of oxygen and an ignition source can ignite suddenly and violently.

_____ 5. The flash point is the lowest temperature at which a liquid produces enough vapor to sustain a continuous fire.

_____ 6. A backdraft can occur when oxygen is introduced into a closed, superheated room.

_____ 7. The three basic ingredients required to create a fire are fuel, oxygen, and air.

_____ 8. Gas has neither independent shape nor volume and tends to expand indefinitely.

_____ 9. A smoke explosion usually occurs when a mixture of flammable gases and oxygen are present in the fire compartment.

_____ 10. Mechanical, electrical, and chemical energy can be converted to heat.

Short Answer

Complete this section with short written answers using the space provided.

1. List the three conditions that must be present for a vapor–air mixture to ignite.

2. List three signs of an impending backdraft.

3. Describe how a flashover can be prevented.

4. Identify the hazards associated with smoke.

5. Identify the four stages of solid fuel fire.

6. Identify the four basic methods of extinguishing fires.

Fire Alarms

The following real case scenarios will give you an opportunity to explore the concerns associated with fire behavior. Read each scenario, and then answer each question in detail.

2. It is 3:00 in the afternoon when your engine company is dispatched to a kitchen fire in a multifamily condominium unit. You and your Lieutenant enter the unit, and it appears that the kitchen has flashed over. You are on the nozzle and your Lieutenant tells you that this is a hot fire and not to disrupt the thermal balance. How should you proceed?

_____
_____
_____
_____
_____

3. Your engine company is dispatched to a two-story, single-family home in a newer development. Upon arrival, you find there is no flame visible and the window glass is smoke-stained with a lot of heat inside. Upon investigation, you see smoke emanating under pressure from cracks. The smoke is puffing and being drawn back like it is breathing. How will you proceed?

_____
_____
_____
_____
_____
_____

Fire Fighter II in Action

The following scenario will give you an opportunity to apply your firefighting knowledge and your fire department SOGs to the new information you learned while studying this chapter. Research your department's SOGs and answer the assignment in detail. Compare your answers with your classmates' and discuss similarities and obvious differences between your answers.

Your company has been dispatched to a reported outside fire at an industrial site. As you respond, you see a red sky from better than ½ mile out. On arrival, you see a lot of fire impinging on a large propane tank. The tank's relief valve has vented and the fire appears to be fed by a propane fill truck (bobcat).

1. What will your size-up report be to dispatch?

2. What resources will you request immediately?

3. What will your initial course of action be?

4. How can you cool the two burning containers in hopes of avoiding a BLEVE?

Building Construction

Workbook Activities

The following activities have been designed to help you. Your instructor may require you to complete some or all of these activities as a regular part of your fire fighter training program. You are encouraged to complete any activity that your instructor does not assign as a way to enhance your learning in the classroom.

Chapter Review

The following exercises provide an opportunity to refresh your knowledge of this chapter.

Matching

Match each of the terms in the left column to the appropriate definition in the right column.

_____ 1. Combustibility   A. A natural material composed of calcium sulfate and water molecules
_____ 2. Thermal conductivity   B. Interior walls extending from the floor to the underside of the floor above
_____ 3. Fire window   C. Built-up unit of construction materials set in mortar
_____ 4. Fire partition   D. Walls designed for structural support
_____ 5. Gypsum   E. The weight of the building contents
_____ 6. Occupancy   F. How a building is used
_____ 7. Live load   G. Describes how readily a material will conduct heat
_____ 8. Load-bearing walls   H. Used when a window is needed in a required fire-resistant wall
_____ 9. Spalling   I. Determines whether a material will burn
_____ 10. Masonry   J. Chipping or pitting of concrete or masonry surfaces

Multiple Choice

Read each item carefully, and then select the best response.

_____ 1. Thermoplastic materials melt and drip when exposed to high temperatures, some even as low as
   A. 100°F (37.8°C).
   B. 250°F (121.1°C).
   C. 500°F (260°C).
   D. 650°F (343.3°C).

_____ 2. What is another term for wood-frame construction?
   A. Type I
   B. Type II
   C. Type IV
   D. Type V

CHAPTER 7

_____ 3. A steel bar joist is an example of a
   A. bowstring truss.
   B. pitched chord truss.
   C. parallel chord truss.
   D. flat chord truss.

_____ 4. When selecting materials for building construction, architects most often place a priority on
   A. price and ease of construction.
   B. functionality and aesthetics.
   C. availability of materials and price.
   D. durability and maintenance expenses.

_____ 5. How many layers will a typical built-up roof covering have?
   A. 3
   B. 5
   C. 7
   D. 9

_____ 6. Which of the following materials will expand at extremely high temperatures, conducts heat well, and loses its strength as the temperature increases?
   A. Steel
   B. Concrete
   C. Masonry
   D. Gypsum

_____ 7. Fire doors and fire windows are rated for a particular duration of
   A. heat resistance to controlled temperatures.
   B. internal temperature compliance.
   C. standard fire resistance.
   D. fire resistance to a standard test fire.

_____ 8. The weight of the building is called the
   A. live load.
   B. total load.
   C. dead load.
   D. structural load.

_____ 9. Which type of glass consists of a thin sheet of plastic between two sheets of glass?
   A. Tempered glass
   B. Wired glass
   C. Laminated glass
   D. Glass blocks

_____ 10. Walls that are constructed on the line between two properties and are shared by a building on each side of the line are called
   A. fire walls.
   B. fire partitions.
   C. curtain walls.
   D. party walls.

_____ 11. The exposed interior surfaces of a building are commonly referred to as the
   A. interior finish.
   B. building surfaces.
   C. structural surfaces.
   D. structural finish.

_____ 12. Which of the following is a commonly used building material?
   A. Steel
   B. Concrete
   C. Aluminum
   D. All of the above

_____ 13. Pitched, curved, and flat are types of
   A. awnings.
   B. roofs.
   C. stairways.
   D. rafters.

_____ 14. Which synthetic material is found in many products and may be transparent or opaque, stiff or flexible, and tough or brittle?
   A. Glass
   B. Plastic
   C. Aluminum
   D. Copper

_____ 15. Trusses are used extensively in support systems for
   A. both floors and roofs.
   B. floors.
   C. roofs.
   D. roofs, with the exception of flat roofs.

_____ 16. What is the length of time that a building or building components can withstand a fire before igniting called?
   A. Pyrolysis
   B. Thermal resistance
   C. Fire retardance
   D. Fire resistance

_____ 17. Thin sheets of wood that are glued together are called
   A. wood panels.
   B. laminated wood.
   C. wood trusses.
   D. wooden beams.

_____ 18. Lightweight and heavy timber construction are examples of
   A. Type I construction.
   B. window frames.
   C. wood floor structures.
   D. roofs.

_____ 19. Which type of building construction has two separate fire loads?
   A. Type II
   B. Type III
   C. Type IV
   D. Type V

_____ 20. Thermoset materials are fused by heat and
   A. will melt in low temperatures.
   B. will not burn.
   C. will always maintain their strength.
   D. will not melt.

_____ 21. Which type of building construction provides the highest degree of safety and is usually made of reinforced concrete and protected steel-frame construction?
- A. Type I
- B. Type II
- C. Type IV
- D. Type V

_____ 22. Buildings having masonry exterior walls, and interior walls, floors, and roofs made of wood, are considered to be
- A. Type II construction.
- B. Type III construction.
- C. Type IV construction.
- D. Type V construction.

Vocabulary

Define the following terms using the space provided.

1. Interior finish:

2. Dead load:

3. Balloon-frame construction:

4. Bowstring truss:

5. Thermoplastic materials:

6. Load-bearing wall:

Fill-in

Read each item carefully, and then complete the statement by filling in the missing word(s).

1. A building with a(n) _____ _____ will have a distinctive curved roof.

2. When wood is exposed to high temperatures, its strength can be decreased through the process of _____.

3. Type _____ building construction is the most commonly used type of construction today.

4. A(n) _____ chord truss is typically used to support a sloping roof.

5. A(n) _____ _____ helps prevent the spread of a fire from one side to the other side of the wall.

6. The term _____ refers to how a building is used.

7. Type _____ is the most fire-resistive category of building construction.

8. Fire severity in a Type II building is determined by the _____ _____ _____ _____.

9. _____-frame construction is used for almost all modern wood-frame construction.

10. The weight of the building's contents is called the _____ _____.

True/False

If you believe the statement to be more true than false, write the letter "T" in the space provided. If you believe the statement to be more false than true, write the letter "F."

_____ 1. The structural components and building contents in Type III, Type IV, and Type V buildings will burn.

_____ 2. Aluminum is more expensive than, and not as strong as, steel.

_____ 3. The entire structure of a manufactured (mobile) home can be destroyed by fire within a few minutes.

_____ 4. Trusses are used in support systems for both floors and roofs.

_____ 5. Fire doors and fire windows are rated for a particular duration of fire resistance to a standard fire test.

_____ 6. Most doors are constructed of aluminum.

_____ 7. Concrete is one of the most commonly used building materials.

_____ 8. Aluminum floors are common in fire-resistive construction.

_____ 9. Type V building construction provides the highest degree of safety.

_____ 10. The support systems for most flat roofs are constructed of aluminum.

Short Answer

Complete this section with short written answers using the space provided.

1. List the five factors that affect how fast wood ignites, burns, and decomposes.

2. Identify and briefly describe the five types of building construction.

3. What problems must be anticipated when considering the fire risks associated with a construction or demolition site?

4. List the four key factors that affect building materials under fire.

5. Briefly describe gusset plates, and how they respond when heated.

6. List the seven major components of a building.

Fundamentals of Fire Fighter Skills

CLUES

Across

2. A roof with sloping or inclined surfaces.
7. Built-up unit of construction or combination of materials such as clay, shale, concrete, glass, gypsum, tile, or stone set in mortar. (NFPA 5000)
9. A roof with a curved shape.
11. Safety glass. There is a thin layer of plastic between two layers of glass, so that the glass does not shatter and fall apart when broken.
13. Buildings constructed since about 1970 that incorporate lightweight construction techniques and engineered wood components. These buildings exhibit less resistance to fire than older buildings.
14. Chipping or pitting of concrete or masonry surfaces. (NFPA 921)

Down

1. A type of safety glass that is heat treated so that, under stress or fire, it will break into small pieces that are not as dangerous.
3. The weight of the aerial device structure and all materials, components, mechanisms, or equipment permanently fastened thereto. (NFPA 1901)
4. A fire-rated assembly used to enclose a vertical opening such as a stairwell, elevator shaft, or chase for building utilities.
5. An older type of construction that used sawn lumber and was built before about 1970.
6. An interior wall extending from the floor to the underside of the floor above.
8. A window assembly rated in accordance with NFPA 257 and installed in accordance with NFPA 80. (NFPA 5000)
10. A horizontal roof, often found on commercial or industrial occupancies.
12. A naturally occurring material consisting of calcium sulfate and water molecules.

Word Fun

The following crossword puzzle is an activity provided to reinforce correct spelling and understanding of terminology associated with firefighting. Use the clues provided to complete the puzzle. Do not include spaces or punctuation when filling in the puzzle.

Fire Alarms

The following real case scenarios will give you an opportunity to explore the concerns associated with building construction. Read each scenario, and then answer each question in detail.

1. Your ladder truck company is dispatched to a structure fire at a three-story Type III construction nursing home. The fire is reported in the kitchen. You start to think about the contents and the building construction. What are your concerns?

   _____
   _____
   _____
   _____
   _____
   _____
   _____

2. It is 10:00 on Friday night when your engine is dispatched to a fire at a bowling alley. Upon arrival, you find a masonry building with a curved roof and the rear exterior wall leaning out. The fire has involved the office area in the rear of the building. How will you proceed?

   _____
   _____
   _____
   _____
   _____
   _____
   _____

Fire Fighter II in Action

The following scenario will give you an opportunity to apply your firefighting knowledge and your fire department SOGs to the new information you learned while studying this chapter. Research your department's SOGs and answer the assignment in detail. Compare your answers with your classmates' and discuss similarities and obvious differences between your answers.

A developer has purchased property in your first response district and is building a new housing development. As you watch the houses take shape, you notice lots of wood trusses, laminated beams, pressed wood I beams, plywood, and OSB being delivered and used.

1. What are your concerns with use of these types of construction material?

   _____
   _____
   _____
   _____
   _____
   _____
   _____

2. How can you prepare to fight fires in this new development?

   _____
   _____
   _____
   _____
   _____
   _____
   _____

Portable Fire Extinguishers

Workbook Activities

The following activities have been designed to help you. Your instructor may require you to complete some or all of these activities as a regular part of your fire fighter training program. You are encouraged to complete any activity that your instructor does not assign as a way to enhance your learning in the classroom.

Chapter Review

The following exercises provide an opportunity to refresh your knowledge of this chapter.

Matching

Match each of the terms in the left column to the appropriate definition in the right column.

_____ 1. Extinguishing agent
_____ 2. Aqueous film-forming foam
_____ 3. Incipient
_____ 4. Hydrostatic testing
_____ 5. Pressure indicator
_____ 6. Ammonium phosphate
_____ 7. FFFP
_____ 8. CO_2
_____ 9. Tamper seal
_____ 10. Fire load

A. Periodic testing of an extinguisher to verify it has sufficient strength to withstand internal pressures
B. Very cold, forms a dense cloud that displaces the air surrounding the fuel
C. An extinguishing agent used on Class B fires that forms a foam layer over the liquid and stops the production of flammable vapors
D. The weight of combustibles in a fire area or on a floor in buildings and structures
E. A retaining device that breaks when the locking mechanism is released
F. The initial stage of a fire
G. An extinguishing agent used in dry chemical fire extinguishers that can be used on Class A, B, and C fires
H. A material used to stop the combustion process
I. Film-forming fluoroprotein foam, a Class B foam additive
J. A gauge on a pressurized portable fire extinguisher that indicates the internal pressure of the expellant

Multiple Choice

Read each item carefully, and then select the best response.

_____ 1. The sodium-chloride-based extinguishing agent that is used in portable fire extinguishers can be
　A. stored in liquid form.
　B. harmful to the environment.
　C. used in all portable fire extinguishers.
　D. applied by hand.

CHAPTER 8

_____ 2. Fire extinguishers weighing more than 40 pounds (18.1 kilograms) should be mounted so that the top of the extinguisher is not more than
   A. 5 feet above the floor.
   B. 3 feet above the floor.
   C. 2 feet above the floor.
   D. 6 feet above the floor.

_____ 3. The best method of transporting a hand-held portable fire extinguisher depends on the
   A. training level of the operator.
   B. size, weight, and design of the extinguisher.
   C. type of extinguishing agent.
   D. size and type of fire.

_____ 4. All fires require
   A. fuel, heat, and oxygen.
   B. fuel and oxygen.
   C. an ignition source.
   D. fuel, heat, oxygen, and carelessness.

_____ 5. Class A fire extinguishers include a number. This number is related to the
   A. type of fuel the fire extinguisher can extinguish.
   B. size of the discharge field.
   C. approximate area of burning fuel the fire extinguisher can extinguish.
   D. amount of water the fire extinguisher holds.

_____ 6. The safest and surest way to extinguish a Class C fire is to turn off the power and
   A. treat it like a Class A or B fire.
   B. treat it like a Class D fire.
   C. treat it like a Class K fire.
   D. treat it like a Class A, B, or D fire.

_____ 7. Class D fires are most often encountered in
   A. kitchens or restaurants.
   B. offices or schools.
   C. machine or repair shops.
   D. hayfields or woodland areas.

_____ 8. Where is the extinguishing agent in a fire extinguisher stored?
   A. Trigger
   B. Nozzle
   C. Cylinder
   D. Handle

_____ 9. Carbon dioxide is a gas that is 1.5 times heavier than
   A. water.
   B. most extinguishing agents.
   C. air.
   D. carbon monoxide.

_____ 10. Two factors to consider when determining the number and types of fire extinguishers that should be placed in each area of occupancy are the
   A. quality and quantities of the fuels.
   B. fuels and ignition sources.
   C. types of fuels and area traffic.
   D. types and quantities of the fuels.

_____ 11. Fires that have not spread past their point of origin are
   A. called introductory fires.
   B. called incipient-stage fires.
   C. easily suppressed.
   D. most often suppressed with an exterior attack.

_____ 12. What is the only dry chemical extinguishing agent rated as suitable for Class A fires?
   A. Potassium chloride
   B. Potassium bicarbonate
   C. Ammonium phosphate
   D. Ammonium bicarbonate

_____ 13. An extinguisher rated 40-B should be able to control a liquid pan fire
   A. with a surface area of 40 ft^2 (3.7 m^2).
   B. within 40 seconds.
   C. 40 times more effectively than a normal Class B extinguisher.
   D. 4 times more effectively than a normal Class B extinguisher.

_____ 14. The three risk classifications according to the amount and type of combustibles that are present in an area are
   A. light, ordinary, and extra hazards.
   B. light, medium, and extra hazards.
   C. normal, light, and extra hazards.
   D. normal, average, and extra hazards.

_____ 15. Class A fires involve
   A. combustible metal fires.
   B. ordinary combustibles.
   C. vegetable oils.
   D. electrically charged materials.

_____ 16. Self-expelling agents do not require
   A. regular maintenance.
   B. tamper seals on the cylinders.
   C. a separate gas cartridge.
   D. maintenance personnel to be specially trained in their use.

_____ 17. Carbon dioxide extinguishers have relatively short discharge ranges of
   A. 1 to 3 feet (0.3 to 0.9 meters).
   B. 3 to 8 feet (0.9 to 2.4 meters).
   C. 10 to 15 feet (3 to 4.6 meters).
   D. 15 to 30 feet (4.6 to 9.1 meters).

_____ 18. Which lever is used to discharge the agent from a portable fire extinguisher?
   A. Trigger
   B. Nozzle
   C. Cylinder
   D. Handle

_____ 19. Vegetable oil fires are classified as
   A. Class A fires.
   B. Class B fires.
   C. Class C fires.
   D. Class K fires.

_____ 20. Class B fire extinguishers can be identified by the
   A. solid red square.
   B. solid blue square.
   C. solid red circle.
   D. solid yellow five-point star.

_____ 21. Carbon dioxide extinguishers are not recommended for
   A. Class B fires.
   B. Class C fires.
   C. outdoor use.
   D. use in kitchens or laboratories.

_____ 22. Electrical rooms should have extinguishers that are approved for use on
   A. Class K fires.
   B. Class A fires.
   C. Class B fires.
   D. Class C fires.

Labeling

Label the following diagram with the correct terms.

1. Basic parts of a portable fire extinguisher.

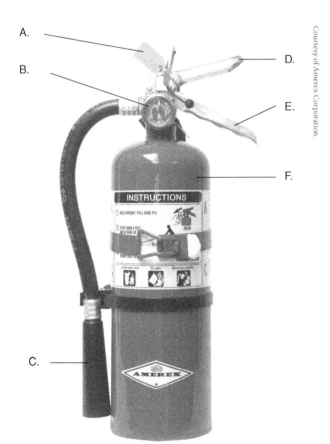

A. _____
B. _____
C. _____
D. _____
E. _____
F. _____

Figure 8-19

Vocabulary

Define the following terms using the space provided.

1. Polar solvent:

2. Extra hazard locations:

3. Extinguishing agent:

4. Cartridge/cylinder fire extinguisher:

5. Underwriters Laboratories, Inc. (UL):

6. Class K fires:

7. Rapid oxidation:

8. Multipurpose dry chemical extinguisher:

Fill-in
Read each item carefully, and then complete the statement by filling in the missing word(s).

1. _____ _____ extinguishers have a short discharge range.

2. The extinguishing agent of a portable extinguisher is discharged through a(n) _____ or horn.

3. An individual with _____ training should be able to use most fire extinguishers effectively.

4. A(n) _____ _____ is a fire extinguishing agent that does not leave a residue when it evaporates.

5. _____ extinguishers are used primarily outdoors for fighting brush and grass fires.

6. Class _____ labels are represented by a solid red square.

7. A Class _____ fire is one that involves wood, cloth, rubber, household rubbish, and some plastics.

8. _____ _____ is a colorless, odorless, electronically nonconductive gas that puts out Class B and C fires by displacing oxygen and cooling the fuel.

9. A fire extinguisher must be _____ after each and every use.

10. The ignition point is the _____ at which a substance will burn.

True/False
If you believe the statement to be more true than false, write the letter "T" in the space provided. If you believe the statement to be more false than true, write the letter "F."

_____ 1. Class K extinguishers are identified by a solid yellow five-point star.

_____ 2. Halon 1211 should be used only when its clean properties are essential.

_____ 3. "Press the trigger" is the first step of PASS.

_____ 4. The bottom of an extinguisher should be mounted at least 4 inches (10.2 centimeters) above the floor.

_____ 5. A Class B extinguisher with a 10-B rating indicates that it is capable of extinguishing the highest level of Class B fires.

_____ 6. Most offices or classrooms would be examples of light hazard areas.

_____ 7. The primary disadvantage of fire extinguishers is their effectiveness.

_____ 8. All fire fighters are trained to perform fire extinguisher maintenance.

_____ 9. Time intervals for testing requirements for an extinguisher are based on construction material and vessel type.

_____ 10. Fire extinguishers can contain several hundred pounds of extinguishing agent.

Short Answer
Complete this section with short written answers using the space provided.

1. Identify the six basic steps in extinguishing a fire with a portable fire extinguisher.

_____
_____
_____
_____
_____
_____

2. Describe the PASS acronym used for fire extinguisher operations.

3. Identify four common indicators that a fire extinguisher needs maintenance.

4. Identify the seven types of fire extinguishers.

5. Identify the six basic parts of most hand-held portable fire extinguishers.

Word Fun

The following crossword puzzle is an activity provided to reinforce correct spelling and understanding of terminology associated with firefighting. Use the clues provided to complete the puzzle. Do not include spaces or punctuation when filling in the puzzle.

CLUES

Across

3 The tapered discharge nozzle of a carbon dioxide–type fire extinguisher.
5 The body of the fire extinguisher where the extinguishing agent is stored.
6 Electrically nonconducting, volatile, or gaseous fire extinguishant that does not leave a residue upon evaporation. (NFPA 2001)
9 A constricting appliance attached to the end of a fire hose or monitor to increase the water velocity and form a stream. (NFPA 1965)
10 The initial or beginning stage of a fire, in which it can be controlled or extinguished by portable extinguishers or small amounts of dry extinguishing agents, without the need for protective clothing or breathing apparatus. (NFPA 484)
11 The process of converting the fatty acids in cooking oils or fats to soap or foam; the action caused by a Class K fire extinguisher.

Down

1 The weight of combustibles in a fire area (measured in ft² or m²) or on a floor in buildings and structures, including either contents or building parts, or both. (NFPA 914)
2 The button or lever used to discharge the agent from a portable fire extinguisher.
4 A chemical process that occurs when a fuel is combined with oxygen, resulting in the formation of ash or other waste products and the release of energy as heat and light.
7 A retaining device that breaks when the locking mechanism is released.
8 The grip used for holding and carrying a portable fire extinguisher.

Fire Alarms

The following real case scenarios will give you an opportunity to explore the concerns associated with portable fire extinguishers. Read each scenario, and then answer each question in detail.

1. It is 7:00 on a Thursday evening when your engine is dispatched to a chimney fire. Upon arrival, you find a two-story, wood-frame residential structure with nothing showing. Upon further investigation, you confirm there is a fire in the chimney. Your Lieutenant tells you to extinguish the fire in the fireplace with an extinguisher. How should you proceed?

   _____
   _____
   _____
   _____
   _____
   _____

2. It is 8:00 on Saturday morning, and your Lieutenant is conducting the morning shift meeting. Saturday's duties are to do a detailed inspection of all equipment. The Lieutenant assigns you to inspect all of the extinguishers on the apparatus and report any that need maintenance. How should you proceed?

   _____
   _____
   _____
   _____
   _____
   _____
   _____
   _____

Fire Fighter II in Action

The following scenario will give you an opportunity to apply your firefighting knowledge and your fire department SOGs to the new information you learned while studying this chapter. Research your department's SOGs and answer the assignment in detail. Compare your answers with your classmates' and discuss similarities and obvious differences between your answers.

You have been dispatched to a kitchen fire, and on arrival you see light smoke showing from a single-family dwelling front door. Your company officer tells you to wait by the engine while he and your partner investigate. He quickly radios back he has food burning on the electric stove and orders you to grab an extinguisher and come to the kitchen.

1. Which extinguisher would you take from the engine, and why?

   _____
   _____
   _____
   _____

2. Your company officer did not mask up when he entered for investigation. Should you mask up before entering, or follow his lead?

   _____
   _____
   _____
   _____

Skill Drills

Skill Drill 8-1: Transporting a Fire Extinguisher
Test your knowledge of this skill drill by filling in the correct words in the photo captions.

1. Locate the closest fire extinguisher.

2. Assess that the extinguisher is _____ and _____ for the type of fire being attacked. Release the _____ _____ _____.

3. Lift the extinguisher using good _____ _____.
   Lift small extinguishers with one hand, and large extinguishers with two hands.

4. Walk _____—do not run—toward the fire. If the extinguisher has a hose and nozzle, carry the extinguisher with one hand and grasp the _____ with the other hand.

Skill Drill 8-2: Attacking a Class A Fire with a Stored-Pressure Water-Type Fire Extinguisher

Test your knowledge of this skill drill by filling in the correct words in the photo captions.

1. Size up the fire to determine whether a(n) _____-_____ water extinguisher is safe and effective for this fire. Ensure that extinguisher is large enough to be safe and effective.

2. Ensure your safety. Make sure you have a(n) _____ _____ from the fire. Do not turn your back on a fire.

3. Remove the _____ and _____. Quickly check the _____ _____ to verify that the extinguisher is adequately charged.

4. Pull the pin to release the extinguisher _____ valve. You must be within 35 to 40 feet (11 to 12 meters) of the fire to be effective.

5. Aim the nozzle and sweep the water stream at the _____ of the flames.

6. Overhaul the fire; take steps to prevent _____, break apart tightly packed fuel, and summon additional help if needed.

Fire Fighter Tools and Equipment

Workbook Activities

The following activities have been designed to help you. Your instructor may require you to complete some or all of these activities as a regular part of your fire fighter training program. You are encouraged to complete any activity that your instructor does not assign as a way to enhance your learning in the classroom.

Chapter Review

The following exercises provide an opportunity to refresh your knowledge of this chapter.

Matching

Match each of the terms in the left column to the appropriate definition in the right column.

_____ 1. Roofman's hook
_____ 2. Multipurpose hook
_____ 3. Size-up
_____ 4. Vertical ventilation
_____ 5. Rapid intervention
_____ 6. Box-end wrench
_____ 7. Sledgehammer
_____ 8. Hydraulic spreader
_____ 9. Battering ram
_____ 10. Thermal imaging device

A. A long pole with a wooden or fiberglass handle and a metal hook on one end used for pulling
B. The process of making openings so that the smoke, heat, and gases can escape vertically from a structure
C. A minimum of two fully equipped personnel on-site, in a ready state, for immediate rescue of injured or trapped fire fighters
D. A long, heavy hammer that requires the use of both hands
E. A long pole with a solid metal hook used for pulling
F. A tool made of hardened steel with handles on the sides used to force doors and to breach walls
G. A hand tool with a closed end used to tighten or loosen bolts
H. A lightweight, hand-operated tool that can produce up to 10,000 pounds of prying and spreading force
I. The observation and evaluation of existing factors that are used to develop objectives, strategy, and tactics for fire suppression
J. Electronic devices that detect differences in temperature based on infrared energy and then generate images based on those data

Multiple Choice

Read each item carefully, and then select the best response.

_____ 1. A pike pole
   A. is a lever used for prying.
   B. is used for rotating.
   C. is a pulling tool.
   D. is a cutting tool.

CHAPTER 9

_____ 2. Which tool heats metal until it melts?
   A. Hydraulic spreader
   B. Power saw
   C. Air bag
   D. Cutting torch

_____ 3. Proper personal protective equipment (PPE) includes
   A. approved helmet.
   B. eye protection.
   C. firefighting gloves.
   D. all of the above.

_____ 4. Tools used for vertical roof ventilation include
   A. shovels and brooms.
   B. power saws and axes.
   C. negative-pressure fans.
   D. rakes and buckets.

_____ 5. Which of the following is not considered part of a fire fighter's PPE?
   A. Boots
   B. Approved firefighting gloves
   C. Approved firefighting prying tool
   D. Personal alert safety system

_____ 6. Which of these are prying tools?
   A. Sledgehammer
   B. Halligan bar and crowbar
   C. K tool and chisel
   D. Bucket and shovel

_____ 7. Pike poles are commonly used for
   A. pulling ceilings.
   B. opening floors.
   C. popping doors off hinges.
   D. car fires.

   8. Which of the following tools is often used in vehicular crashes to gain access to a victim who needs care?
   A. Spring-loaded center punch
   B. Sledgehammer
   C. Chainsaw
   D. Crowbar

_____ 9. After use, all hand tools should be completely cleaned and
   A. scientifically tested.
   B. inspected.
   C. sharpened.
   D. placed in the tool cabinet.

____ 10. Which of the following is considered a tool for cutting metal?
   A. Crowbar
   B. Drywall hook
   C. Pick-head axe
   D. Hacksaw

____ 11. Which of the following is considered a tool for rotating?
   A. Claw bar
   B. Ceiling hook
   C. Axe
   D. Screwdriver

____ 12. Wood handles on tools should be
   A. sanded and painted.
   B. sanded and varnished.
   C. sanded and linseed oil applied.
   D. left alone.

____ 13. Which of the following is a basic piece of equipment for interior firefighting?
   A. Hand light or portable light
   B. Thermal imaging device
   C. Chain saw
   D. Exhaust fan

____ 14. Which tool is used to cut chains or padlocks?
   A. Bolt cutter
   B. Battering ram
   C. Flat-head axe
   D. K tool

____ 15. Which of the following is not a mechanical saw?
   A. Chain saw
   B. Rotary saw
   C. Reciprocating saw
   D. Hacksaw

____ 16. Which of the following is considered a tool for striking?
   A. Crowbar
   B. Drywall hook
   C. Pick-head axe
   D. Hacksaw

____ 17. Special equipment to be carried by a rapid intervention crew (RIC) should include
   A. a thermal imager.
   B. prying tools.
   C. striking tools.
   D. all of the above.

____ 18. At a single-family residential house fire, tool staging should be located
   A. at the first-due firehouse.
   B. outside the building.
   C. inside the structure.
   D. on the second ladder company.

____ 19. Which of the following is not a basic search and rescue hand tool?
   A. Halligan tool
   B. Axe
   C. Hand light
   D. K tool

____ 20. All power equipment should be left in a ready state, which includes
   A. fuel tanks filled completely with fresh fuel.
   B. hydraulic hoses, if applicable, cleaned and inspected.
   C. the removal and replacement of any dull or damaged blades.
   D. all of the above.

Vocabulary

Define the following terms using the space provided.

1. Claw bar:

2. Reciprocating saw:

3. Overhaul:

4. Gripping pliers:

5. Crowbar:

6. Seat belt cutter:

7. Spanner wrench:

8. Kelly tool:

9. Cutting torch:
_____
_____
_____

10. Hydrant wrench:
_____
_____
_____

11. Ceiling hook:
_____
_____
_____

12. Pike pole:
_____
_____
_____

Fill-in
Read each item carefully, and then complete the statement by filling in the missing word(s).

1. A fire fighter must know how to use tools _____, efficiently, and safely.

2. _____ is the phase during which fire fighters start thinking about the possible tools or equipment they may need during an incident.

3. _____ is a prime consideration when using any tools and equipment.

4. A(n) _____ is a specialized striking tool with an axe on one end of the head and a sledgehammer on the other end.

5. _____ or _____ tools allow a fire fighter to increase the power exerted upon an object and extend the fire fighter's reach.

6. A(n) _____-_____ _____ has a pointed "pick" on one end of the head and an axe blade on the other end.

7. To reduce the total number of tools needed to achieve a goal, a fire fighter may carry a tool that has a number of uses. This tool is categorized as a(n) _____-_____ tool.

8. _____ _____ are tools that use extremely high-temperature flames to cut through an object.

9. During an incident, the tools or equipment that may be required are often placed in a designated area referred to as the _____ _____ area.

10. The _____-_____ manuals provide recommendations and instructions on how to clean, care for, and inspect tools and equipment.

True/False
If you believe the statement to be more true than false, write the letter "T" in the space provided. If you believe the statement to be more false than true, write the letter "F."

_____ 1. Horizontal ventilation may be achieved by opening a window.

_____ 2. Vertical ventilation is always achieved by opening windows.

_____ 3. Department guidelines or standard operating procedures usually guide the decision for which tool to use during an incident.

_____ 4. Air bags can be used to lift heavy objects.

_____ 5. Striking tools should be assigned to crews only during the forcible entry phase of a response.

_____ 6. The K tool is used to pull the lock cylinder out of a door.

_____ 7. Interior attack can be orchestrated by any response member at any time during an emergency response.

_____ 8. Mechanically powered equipment is more powerful than manually powered equipment.

_____ 9. New fire fighters are often surprised by the strength and energy required to perform many tasks.

_____ 10. An RIC should carry self-contained breathing apparatus (SCBA) and spare air cylinders.

Short Answer

Complete this section with short written answers using the space provided.

1. What are two advantages of using pushing/pulling tools? Give five examples of pushing/pulling tools.

2. Why is it important to know which tools are needed for each phase of an incident?

3. Identify the tools used during overhaul operations.

4. Identify the basic set of tools used for interior firefighting.

5. Describe the general functions of the six categories of tools, and provide an example from each category.

6. List the components of a full set of PPE.

7. List the special equipment needed for ventilation.

8. Identify the basic set of tools used for search and rescue.

9. Identify the tools and special equipment the RIC should carry and have ready for immediate response.

10. Explain the importance of tool and equipment maintenance for a fire fighter.

Fundamentals of Fire Fighter Skills

CLUES

Across

5. A power saw that uses the rotating movement of a chain equipped with sharpened cutting edges. It is typically used to cut through wood.
7. A multipurpose tool that can be used for several forcible entry and ventilation applications because of its unique head design.
8. A tool with a pointed claw-hook on one end and a forked- or flat-chisel pry on the other end. It is often used for forcible entry.
9. A long pole with a solid metal hook used for pulling.
13. A striking tool.
14. A long pole with a pointed head and two retractable cutting blades on the side.
15. The observation and evaluation of existing factors that are used to develop objectives, strategy, and tactics for fire suppression. (NFPA 1051)
16. A specialized striking tool, weighing 6 pounds (3 kilograms) or more, with an axe on one end and a sledgehammer on the other end.

Down

1. A metal tool with one sharpened end that is used to break apart material in conjunction with a hammer, mallet, or sledgehammer.
2. A wrench having one fixed grip and one movable grip that can be adjusted to fit securely around pipes and other tubular objects.
3. A tool that is used to remove lock cylinders from structural doors so the locking mechanism can be unlocked.
4. A cutting tool designed for use on metal. Different blades can be used for cutting different types of metal.
6. A long, heavy hammer that requires the use of both hands.
10. The process of final extinguishment after the main body of a fire has been knocked down. All traces of fire must be extinguished at this time. (NFPA 402, 2002)
11. A combination of tools, usually consisting of a Halligan tool and a flat-head axe, that are commonly used for forcible entry.
12. A short-handled hammer.

Word Fun

The following crossword puzzle is an activity provided to reinforce correct spelling and understanding of terminology associated with firefighting. Use the clues provided to complete the puzzle. Do not include spaces or punctuation when filling in the puzzle.

Fire Alarms

The following real case scenarios will give you an opportunity to explore the concerns associated with fire fighter tools and equipment. Read each scenario, and then answer each question in detail.

1. It is 6:00 on a Monday morning when your engine is dispatched to a commercial structure fire in a warehouse. You are the fifth engine to arrive at the scene. The IC assigns your engine to the RIC group. Your Lieutenant tells you to gather the RIC equipment and place it in the staging area. How will you proceed?

   _____
   _____
   _____
   _____
   _____
   _____

2. It is 7:30 on a Sunday morning and your Lieutenant is conducting a shift meeting. The Lieutenant tells you to ensure the power tools and equipment are clean and inspected. After the meeting, you start with the ladder truck by pulling off the chainsaws. How should you proceed?

   _____
   _____
   _____
   _____
   _____
   _____

Fire Fighter II in Action

The following scenario will give you an opportunity to apply your firefighting knowledge and your fire department SOGs to the new information you learned while studying this chapter. Research your department's SOGs and answer the assignment in detail. Compare your answers with your classmates' and discuss similarities and obvious differences between your answers.

Power tools and equipment are used for lighting, ventilation, salvage, and overhaul. They are used extensively in the fire service. Without proper training and maintenance, these tools can fail to perform properly or perform at all. They are also very dangerous when used by untrained personnel.

1. Make a list of the frequent maintenance that should be performed on the power tools carried by your company.

   _____
   _____
   _____
   _____
   _____
   _____

2. What steps will you take to ensure these tools are ready after each use?

   _____
   _____
   _____
   _____
   _____
   _____

Ropes and Knots

Workbook Activities

The following activities have been designed to help you. Your instructor may require you to complete some or all of these activities as a regular part of your fire fighter training program. You are encouraged to complete any activity that your instructor does not assign as a way to enhance your learning in the classroom.

Chapter Review

The following exercises provide an opportunity to refresh your knowledge of this chapter.

Matching

Match each of the terms in the left column to the appropriate definition in the right column.

_____ 1. Safety knot
_____ 2. Twisted rope
_____ 3. Standing part
_____ 4. Hitches
_____ 5. Block creel construction
_____ 6. Dynamic rope
_____ 7. Escape rope
_____ 8. Bight
_____ 9. Static rope
_____ 10. Carabiner

A. The part of a rope between the working end and the running end
B. Rope constructed without knots or splices in the yarns, ply yarns, strands, braids, or rope
C. An oval-shaped device with a self-closing gate that can be used for connecting pieces of rope, webbing, or other hardware
D. A U-shape created by bending a rope with the two sides parallel
E. A knot used to secure the leftover working end of the rope; also known as an overhand knot or keeper knot
F. A rope generally made out of synthetic material that stretches very little under load
G. Rope constructed of fibers twisted into strands, which are then twisted together
H. An emergency use rope designed to carry the weight of only one person and to be used only once
I. Knots that wrap around an object
J. A rope generally made out of synthetic materials that is designed to be elastic and stretch when loaded

Multiple Choice

Read each item carefully, and then select the best response.

_____ 1. What is the working end of the rope used for?
   A. Lifting or hoisting
   B. Securing the knot
   C. Carrying the rope
   D. Forming the knot

CHAPTER 10

_____ 2. What are three primary types of rope used in the fire service?
   A. Polypropylene, synthetic, and nylon
   B. Natural fiber, manila, and sisal
   C. Life safety, utility, and escape
   D. Hauling, securing, and towing

_____ 3. The running end is the part of the rope used for
   A. forming the knot.
   B. life safety.
   C. securing the knot.
   D. lifting or hoisting.

_____ 4. The knot used to secure the leftover working end of a rope is called a
   A. granny knot and severe tangle.
   B. becket bend.
   C. slip knot.
   D. safety knot.

_____ 5. Natural fiber ropes
   A. lose their load-carrying ability over time.
   B. absorb water.
   C. degrade quickly.
   D. all of the above.

_____ 6. A pike pole is hoisted using
   A. a figure eight and a clove hitch.
   B. clove and half hitches.
   C. a figure eight on a bight.
   D. a safety knot and bowline.

_____ 7. Life safety rope is used solely for
   A. sectioning off safety areas.
   B. supporting people.
   C. medical response.
   D. hoisting equipment.

_____ 8. Synthetic ropes
   A. absorb more water than natural fiber ropes.
   B. cannot be used for life safety rope.
   C. can be damaged by ultraviolet light.
   D. will not be damaged by acids.

_____ 9. Fire department ropes may be cleaned by using
   A. mild alkali solution.
   B. mild acid solution.
   C. mild soap and water.
   D. window cleaner.

_____ 10. Life safety rope is made of
   A. continuous filament virgin fiber and woven of block creel construction.
   B. the strongest natural fiber available.
   C. lightweight water-resistant fibers.
   D. fibers tested by the NFPA.

_____ 11. Polypropylene rope is often used in water rescue because
   A. it is light, does not absorb water, and floats.
   B. it is less expensive than natural rope.
   C. it is light and has built-in water repellents.
   D. it is easier to control.

_____ 12. The load-carrying capacity or strength of the rope will be
   A. increased by any knot.
   B. reduced by any knot.
   C. maintained by any knot.
   D. maintained by a bend or hitch.

_____ 13. Utility ropes
   A. may be used for a personal safety rope.
   B. can bear a single person (300 pounds [136.1 kilograms]).
   C. can bear two persons (600 pounds [272.2 kilograms]).
   D. are not to be used as a life safety rope.

_____ 14. How is a loop formed?
   A. By using the standing part of the rope
   B. By rolling from the end of the rope
   C. By making a circle in the rope
   D. By using the standing part of the knot

_____ 15. When hoisting an axe, a good rule of thumb is to
   A. hoist the equipment in a vertical position.
   B. use one-person rope.
   C. use dynamic rope.
   D. execute the hoist as quickly as possible.

_____ 16. Life safety ropes are rated under the minimum requirements set by
   A. NRA.
   B. NFPA.
   C. NIOSH.
   D. CDC.

_____ 17. Which device is used to connect one rope to another?
   A. Carabiner
   B. Kernmantle
   C. Harness
   D. Stokes

Labeling

Label the following diagram with the correct terms.

1. Sections of a rope used in tying knots.

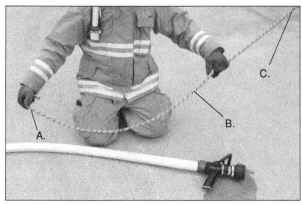

A. _____
B. _____
C. _____

© Jones & Bartlett Learning. Photographed by Glen E. Ellman.

Figure 10-23

Vocabulary

Define the following terms using the space provided.

1. Running end:

2. Knot:

3. Braided rope:

4. Rope bag:

5. Depressions:

6. Shock load:

7. Kernmantle rope:

8. Working end:

9. Round turn:

10. Harness:

Fill-in
Read each item carefully, and then complete the statement by filling in the missing word(s).

1. _____ are used to fasten rope or webbing to objects or to each other.
2. A(n) _____ is formed by reversing the direction of the rope to form a U bend with two parallel ends.
3. _____ are knots that wrap around an object such as a pike pole or fencepost.
4. Any knot will _____ the load-carrying capacity or strength of the rope.
5. A(n) _____ _____ is used to secure the leftover working end of the rope to the standing part of the rope.
6. The figure eight on a bight knot creates a _____ _____ at the working end of a rope.
7. A(n) _____ _____ is formed by making a loop and then bringing the two ends of the rope parallel to each other.
8. Each piece of rope must have a(n) _____ _____ that details its history, usage, type of use, and loads applied.
9. A(n) _____ is a piece of rescue or safety equipment worn by a person and used to secure that person to a rope or a solid object.
10. A _____ is commonly used to connect one rope to another rope, a harness, or itself.

True/False
If you believe the statement to be more true than false, write the letter "T" in the space provided. If you believe the statement to be more false than true, write the letter "F."

_____ 1. Escape ropes can be used only once.
_____ 2. All knots require a safety knot for completion.
_____ 3. Kernmantle ropes are the only ropes subject to being shock-loaded.
_____ 4. Hitches are used to attach a rope around an object.
_____ 5. Life safety ropes must be inspected after each use.
_____ 6. Life safety ropes can be made of natural fibers if properly inspected, recorded, and maintained.
_____ 7. All knots reduce the load-carrying capacity of a rope.
_____ 8. A static rope is better suited for most fire rescue situations.

Short Answer

Complete this section with short written answers using the space provided.

1. Identify the four parts of the rope maintenance formula.

2. List and describe the three types of rope construction.

3. List and describe eight simple knots and their usage in the fire service.

4. List four steps recommended for cleaning ropes.

5. List and describe the most common synthetic fiber ropes used for fire department operations.

6. What are some of the drawbacks of using natural fiber ropes?

7. List the principles to preserve the strength and integrity of rope.

8. What are some of the advantages of using synthetic fiber ropes?

Fundamentals of Fire Fighter Skills

CLUES

Across

1. Rope used on extension ladders to raise a fly section.
4. The part of a rope used for lifting or hoisting.
8. A knot that attaches to or wraps around an object so that when the object is removed, the knot will fall apart. (NFPA 1670)
9. An auxiliary equipment system item; load-bearing connector with a self-closing gate used to join other components of life safety rope. (NFPA 1983)
10. Indentations felt on a kernmantle rope that indicate damage to the interior (kern) of the rope.
13. An instantaneous load that places a rope under extreme tension, such as when a falling load is suddenly stopped as the rope becomes taut.
14. Rope constructed of fibers twisted into strands, which are then twisted together.

Down

2. A rope generally made from synthetic materials and that is designed to be elastic and stretch when loaded. It is often used by mountain climbers.
3. A U shape created by bending a rope with the two sides parallel.
4. A bag used to protect and store rope so that the rope can be easily and rapidly deployed without kinking.
5. A knot used to join the ends of webbing together.
6. Rope constructed by intertwining strands in the same way that hair is braided.
7. A knot used to join two ropes together.
8. A piece of equipment worn by a rescuer that can be attached to a life safety rope.
11. A fastening made by tying together lengths of rope or webbing in a prescribed way. (NFPA 1670)
12. A piece of rope formed into a circle.

Word Fun

The following crossword puzzle is an activity provided to reinforce correct spelling and understanding of terminology associated with firefighting. Use the clues provided to complete the puzzle. Do not include spaces or punctuation when filling in the puzzle.

Fire Alarms

The following real case scenarios will give you an opportunity to explore the concerns associated with ropes and knots. Read each scenario, and then answer each question in detail.

1. It is 9:00 on a Thursday evening when your engine company is dispatched to a structure fire at a three-story townhouse. Upon arrival, you are assigned to staging. The fire has been extinguished on the third floor. The IC assigns you and your partner to retrieve a ventilation fan from your engine and secure it to the rope that is deployed from the third floor. How should you proceed?

   _____
   _____
   _____
   _____
   _____
   _____
   _____

2. You have just returned from an apartment fire. During the fire, you used some of the utility rope off your engine to raise and lower equipment from the roof. While using the rope, it became dirty and needs to be cleaned and inspected. How will you proceed?

   _____
   _____
   _____
   _____
   _____
   _____
   _____

Fire Fighter II in Action

The following scenario will give you an opportunity to apply your firefighting knowledge and your fire department SOGs to the new information you learned while studying this chapter. Research your department's SOGs and answer the assignment in detail. Compare your answers with your classmates' and discuss similarities and obvious differences between your answers.

You have just finished up a six-hour Introduction to Rope class. You have been introduced to different types of rope, several knots, and hitches. After returning to quarters you ask your company officer when you would ever be required to use this new knowledge.

Your company officer orders you to do a bit of research and answer your own question. He also gives you a 30-minute time window, until the next work shift, to make your presentation.

1. Develop an outline addressing the different applications for which the fire service has used rope.

   _____
   _____
   _____
   _____
   _____
   _____
   _____

Skill Drills

Skill Drill 10-5: Tying a Safety Knot
Test your knowledge of this skill drill by filling in the correct words in the photo captions.

1. Take the _____ end of the rope, beyond the _____, and form a loop around the _____ part of the rope.

2. Pass the loose end of the rope through the _____.

3. Tighten the safety knot by pulling on _____ _____ at the same time.

4. Test whether you have tied a safety knot correctly by sliding it on the _____ part of the rope. A knot that is tied correctly will _____.

Skill Drill 10-10: Tying a Figure Eight on a Bight

Test your knowledge of this skill drill by filling in the correct words in the photo captions.

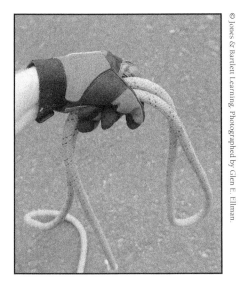

1. Form a(n) _____ and identify the _____ end of the bight as the _____ _____ of the rope.

2. Holding both sides of the bight together, form a(n) _____.

3. Feed the _____ _____ of the bight back through the _____.

4. Pull the _____ tight.

5. Secure the loose end of the rope with a(n) _____ knot.

Skill Drill 10-13: Tying a Bowline

Test your knowledge of this skill drill by placing the photos below in the correct order. Number the first step with a "1," the second step with a "2," and so on.

_____ Form another small loop in the standing part of the rope with the section close to the working end on top. Thread the working end up through this loop from the bottom.

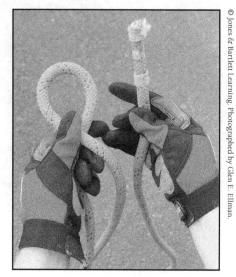

_____ Make the desired sized loop and bring the working end back to the standing part.

_____ Tighten the knot by holding the working end and pulling the standard part of the rope backward.

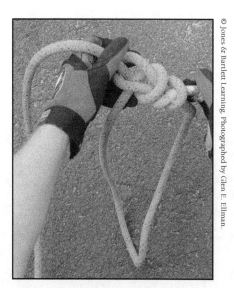

_____ Tie a safety knot in the working end of the rope.

_____ Pass the working end over the loop, around and under the standing part, and back down through the same opening.

Skill Drill 10-19: Hoisting a Charged Hose Line
Test your knowledge of this skill drill by filling in the correct words in the photo captions.

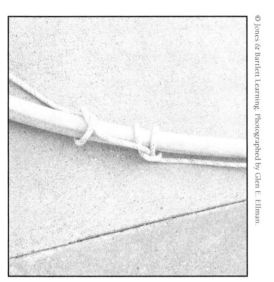

1. Make sure that the nozzle is completely _____ and secure. Use a _____ _____, 1 or 2 feet (0.3–0.6 meters) behind the nozzle, to tie the end of the _____ _____ around a charged hose line. Use a(n) _____ _____ to secure the loose end of the rope below the _____ _____.

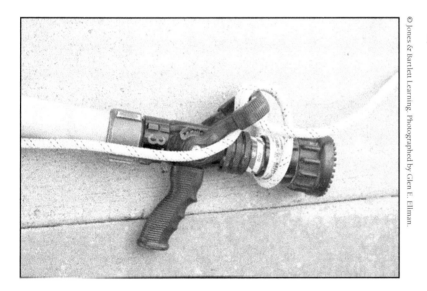

2. Make a(n) _____ in the rope even with the nozzle shut-off handle. Insert the _____ through the handle opening and slip it over the end of the _____. When the bight is pulled tight, it will create a _____ hitch and secure the handle in the _____ position while the charged hose line is hoisted. Communicate with the fire fighter above that the hose line is ready to hoist.

Response and Size-Up

Workbook Activities

The following activities have been designed to help you. Your instructor may require you to complete some or all of these activities as a regular part of your fire fighter training program. You are encouraged to complete any activity that your instructor does not assign as a way to enhance your learning in the classroom.

Chapter Review

The following exercises provide an opportunity to refresh your knowledge of this chapter.

Matching

Match each of the terms in the left column to the appropriate definition in the right column.

_____ 1. Size-up
_____ 2. Personnel accountability system
_____ 3. Preincident plan
_____ 4. Salvage
_____ 5. Freelancing
_____ 6. Extension
_____ 7. Resources
_____ 8. Exposure
_____ 9. Response
_____ 10. Overhaul

A. Taking action on your own without regard to standard operating procedures, command structure, or the strategic plan
B. The movement of fire into uninvolved areas of a structure
C. The ongoing observation and evaluation of factors that are used to develop objectives, strategy, and tactics for fire suppression
D. All of the means available to fight a fire or conduct emergency operations
E. Activities that occur in preparation for an emergency and continue until the arrival of emergency apparatus at the scene
F. Removing or protecting property that could be damaged during firefighting or overhaul operations
G. Examination of the building and contents involved in a fire to ensure that the fire is completely extinguished
H. Provides details about a building's construction, layout, contents, special hazards, and fire protection systems
I. An updated list of fire fighters assigned to each vehicle or crew
J. An area adjacent to the fire that may become involved if not protected from a fire

Multiple Choice

Read each item carefully, and then select the best response.

_____ 1. A preincident plan is helpful during size-up because it contains
   A. information about the structure.
   B. the potential number of units needed for response.
   C. the potential equipment requirements for a response.
   D. information about the weather.

CHAPTER 11

_____ 2. The gas supply to a building is usually controlled by
   A. a single valve at the entry point of the gas piping.
   B. an underground valve that requires the use of a special wrench.
   C. qualified technicians.
   D. the property owner.

_____ 3. Dispatch information will include
   A. the location of the incident and the type of attack required.
   B. the location of the incident and initial scene assessment.
   C. the location of the incident, the type of emergency, and the units due to respond.
   D. the type of emergency, scene assessment, and initial scene assessment.

_____ 4. The incident commander (IC) assembles, interprets, and bases decisions on information presented in the
   A. call-out.
   B. size-up.
   C. reconnaissance report.
   D. dispatch message.

_____ 5. If an incident requires more resources than the local community can provide, most departments have
   A. relief workers.
   B. agreements with state or provincial training institutions.
   C. mutual aid agreements.
   D. support response teams.

_____ 6. Fire fighters who respond to an incident on the fire apparatus deposit their personnel accountability tags
   A. in the command post.
   B. with the incident commander.
   C. with the rapid intervention team.
   D. on a designated location on the apparatus.

_____ 7. When fire fighters advance into the fire building with hose lines to overpower the fire, they are part of a(n)
   A. defensive attack.
   B. defensive response.
   C. offensive attack.
   D. rapid intervention team.

_____ 8. An emergency vehicle must always be operated with
   A. due regard for the safety of everyone on the road.
   B. the assurance that all drivers will yield to the emergency vehicle's right of way.
   C. the skills learned in training.
   D. the intention to arrive on scene as soon as possible.

_____ 9. The removal or protection of property that could be damaged during firefighting is called
   A. postincident reporting.
   B. overhaul.
   C. recovery.
   D. salvage.

_____ 10. Why is the age of the building often an important consideration in size-up?
   A. Older buildings burn faster.
   B. Building and fire safety codes change over time.
   C. Newer buildings have higher property values.
   D. Ventilation is often easier to perform on new homes.

_____ 11. During the response phase, the fire fighter should begin to
   A. consider any factors that could affect the situation.
   B. rest and prepare for the upcoming incident.
   C. have constant communications with the driver.
   D. organize equipment.

_____ 12. The main area of the fire is the
   A. hot spot.
   B. seat of the fire.
   C. attack area.
   D. target area.

_____ 13. To track all fire fighters there should be a(n) _____ at every incident scene.
   A. incident commander
   B. personnel accountability system
   C. incident management system
   D. accountability officer

_____ 14. Events and outcomes that can be predicted based on facts, observations, common sense, and previous experience are called
   A. estimates.
   B. report items.
   C. part of the size-up.
   D. probabilities.

_____ 15. A fire department's basic resources are
   A. its personnel and apparatus.
   B. its preincident plans and trained personnel.
   C. its specialized equipment and apparatus.
   D. its specially trained personnel.

_____ 16. What is the first consideration at any emergency incident?
   A. Protecting property
   B. Protecting lives
   C. Controlling traffic
   D. Completing a full size-up

_____ 17. What is the secondary objective at any emergency incident?
   A. Protecting property
   B. Protecting lives
   C. Protecting fire fighters
   D. Protecting bystanders

_____ 18. Who develops the incident action plan that outlines the steps needed to control the situation?
   A. The initial attack team
   B. The dispatcher
   C. The incident management system
   D. The incident commander

_____ 19. The initial size-up of an incident is conducted
   A. by the bystanders on scene.
   B. by the first officers on scene.
   C. when the first unit arrives on scene.
   D. by reviewing the preincident plan.

_____ 20. During an incident, shutting off electrical service eliminates potential
   A. damage to fire department equipment.
   B. interference with communications equipment.
   C. ignition sources.
   D. structural obstructions.

Vocabulary

Define the following terms using the space provided.

1. Personnel accountability tag (PAT):

2. Size-up:

3. Thermal imaging devices:

4. Freelancing:

5. Response actions:

Fill-in
Read each item carefully, and then complete the statement by filling in the missing word(s).

1. _____ are events that can be predicted or anticipated, based on facts, observations, and previous experiences.

2. After a fire is under control, a(n) _____ is conducted to completely extinguish any remaining fires.

3. The _____ report is created for the incident commander based on the inspection and exploration of a specific area.

4. The size-up process requires a(n) _____ approach to managing information.

5. All equipment should be properly mounted, stowed, or _____ on the fire apparatus to prevent injury.

6. The shut-off valve for a natural gas system is usually a(n) _____ -turn valve with a locking device.

7. A(n) _____ attack occurs when fire fighters advance into the fire with extinguishing agents to overpower the fire.

8. The _____-_____ information is essential for determining the appropriate strategy and tactics for an incident.

9. Older wooden buildings may have utilized _____-_____ construction, which can provide a path for fire to spread rapidly into uninvolved areas.

10. The IC who adopts a(n) _____ strategy has determined that there is no property left to save or that the potential for saving property does not justify the risk to fire fighters.

True/False
If you believe the statement to be more true than false, write the letter "T" in the space provided. If you believe the statement to be more false than true, write the letter "F."

_____ 1. Defensive strategies are used to protect exposed properties because defensive strategies are ineffective in extinguishing fires.

_____ 2. Shutting off water service in a residential fire is often difficult because of the number of valves that need to be closed within the structure.

_____ 3. Fire fighters should not respond to an emergency incident unless they are dispatched.

_____ 4. A fire fighter can make a better size-up by standing up in the apparatus while approaching the scene of a working fire.

_____ 5. Visible flames provide all of the information needed for the incident commander to make decisions on where to set up the attack teams.

_____ 6. Traffic is often a major concern for fire fighter safety at an emergency incident.

_____ 7. Never attempt to mount fire apparatus while it is moving.

_____ 8. The driver of the fire apparatus is legally responsible for the safe operation of the apparatus at all times.

_____ 9. Freelancing is dangerous.

_____ 10. During transport, it is important to have constant contact with the driver to make him or her aware of all response information.

Short Answer

Complete this section with short written answers using the space provided.

1. Why is it important for fire fighters to understand how size-up is completed?

2. In order of priority, identify the five basic fire-ground objectives.

3. Identify why the three main utilities need to be controlled at an emergency incident.

4. Identify and describe the two basic categories of information utilized in size-up.

108 Fundamentals of Fire Fighter Skills

CLUES

Across

1. The observation and evaluation of existing factors that are used to develop objectives, strategy, and tactics for fire suppression. (NFPA 1051)
6. A return to flaming combustion after apparent but incomplete extinguishment. (NFPA 921)
7. Fire that moves into areas not originally involved in the incident, including walls, ceilings, and attic spaces; also, the movement of fire into uninvolved areas of a structure.
9. The process of final extinguishment after the main body of a fire has been knocked down. All traces of fire must be extinguished at this time. (NFPA 402)
11. The deployment of an emergency service resource to an incident. (NFPA 901)
12. The main area of the fire.
13. An identification card used to track the location of a fire fighter on an emergency incident.

Down

2. A written document resulting from the gathering of general and detailed information to be used by public emergency response agencies and private industry for determining the response to reasonable anticipated emergency incidents at a specific facility.
3. An advance into the fire building by fire fighters with hose lines or other extinguishing agents that are intended to overpower the fire.
4. The dangerous practice of acting independently of command instructions.
5. An older type of wood frame construction in which the wall studs extend vertically from the basement of a structure to the roof without any fire stops.
8. A firefighting procedure for protecting property from further loss following an aircraft accident or fire. (NFPA 402)
10. The heat effect from an external fire that might cause ignition of, or damage to, an exposed building or its contents. (NFPA 80A)

Word Fun

The following crossword puzzle is an activity provided to reinforce correct spelling and understanding of terminology associated with firefighting. Use the clues provided to complete the puzzle. Do not include spaces or punctuation when filling in the puzzle.

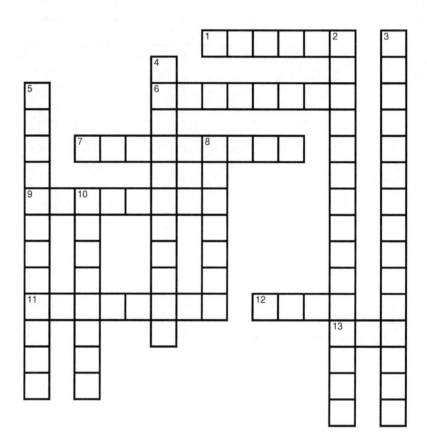

Fire Alarms

The following real case scenarios will give you an opportunity to explore the concerns associated with response and size-up. Read each scenario, and then answer each question in detail.

1. You are about to eat dinner when you are dispatched to a vehicle fire. It is dusk when you arrive, and it is rush hour. Your Lieutenant orders everyone to put on their high-visibility traffic vests and tells you to put out warning devices to help secure the scene. How should you proceed?

   _____
   _____
   _____
   _____
   _____
   _____

2. Your engine company is dispatched to a commercial structure fire in a large pizza restaurant. The building is a wood-frame construction with a lightweight truss. The building is 75 percent involved with fire in the attic area. You are the acting officer and need to decide the overall strategy. How will you proceed?

   _____
   _____
   _____
   _____
   _____
   _____

Fire Fighter II in Action

The following scenario will give you an opportunity to apply your firefighting knowledge and your fire department SOGs to the new information you learned while studying this chapter. Research your department's SOGs and answer the assignment in detail. Compare your answers with your classmates' and discuss similarities and obvious differences between your answers.

Your ladder company has responded to a structure fire in a single family dwelling, and you have been assigned the exterior work. This includes laddering the structure and shutting off all utilities.

1. Explain how to shut off the electricity to a typical single family dwelling in your jurisdiction.

2. Explain how to shut off the gas service to a typical single family dwelling in your jurisdiction.

3. Explain how to shut off the water service to a typical single family dwelling in your jurisdiction.

Skill Drills

Skill Drill 11-1: Mounting Apparatus
Test your knowledge of the skill drill by filling in the correct words in the photo captions.

1. When mounting (climbing aboard) fire apparatus, always have at least one hand firmly grasping a(n) _____ and at least one foot firmly placed on a foot surface. Maintain the one hand and one foot placement until you are _____.

2. Fasten your seat belt and then don any other required safety equipment for response, such as _____ _____ and intercom.

Forcible Entry

Workbook Activities

The following activities have been designed to help you. Your instructor may require you to complete some or all of these activities as a regular part of your fire fighter training program. You are encouraged to complete any activity that your instructor does not assign as a way to enhance your learning in the classroom.

Chapter Review

The following exercises provide an opportunity to refresh your knowledge of this chapter.

Matching

Match each of the terms in the left column to the appropriate definition in the right column.

_____ 1. Interior wall   A. A door design that consists of wood core blocks inside the door
_____ 2. Forcible entry   B. Usually the best point for forcing entry into a vehicle or structure
_____ 3. Halligan   C. One part of the forcible entry tool "irons"
_____ 4. Latch   D. The U-shaped part of a padlock
_____ 5. Bite   E. A small opening made to enable better tool access in forcible entry
_____ 6. Solid-core   F. Gaining access to a structure when the normal means of entry are locked, secured, obstructed, blocked, or unable to be used
_____ 7. Rabbet   G. A cutting tool with a pry bar built into the cutting part of the tool
_____ 8. A tool   H. A type of door frame that has the stop for the door cut into the frame
_____ 9. Shackles   I. The part of the door lock that catches and holds the door frame
_____ 10. Door   J. A wall inside a building that divides a large space into smaller areas

Multiple Choice

Read each item carefully, and then select the best response.

_____ 1. Which type of tool is designed to cut into a lock cylinder and has a pry bar built into the cutting part of the tool?
   A. A tool
   B. K tool
   C. J tool
   D. Adz

_____ 2. Walls that support the rafters and/or ceiling of a structure are
   A. nonbearing.
   B. exterior walls.
   C. load bearing.
   D. partitions.

CHAPTER 12

_____ 3. The main part of a padlock that houses the locking mechanisms is the
   A. lock body.
   B. shackle.
   C. unlocking device.
   D. deadbolt.

_____ 4. Doors that have two sections and a double track where one side is fixed and the other slides are known as
   A. sliding doors.
   B. tempered doors.
   C. slab doors.
   D. honeycomb doors.

_____ 5. The two door locks that can be surface mounted on the interior of the door frame are rim locks and
   A. mortise locks.
   B. combination locks.
   C. key locks.
   D. deadbolts.

_____ 6. Gaining access to a structure when the normal means of entry are unable to be used is referred to as
   A. structure entry.
   B. forcible entry.
   C. operating access.
   D. forced access.

_____ 7. Vehicle windshields are most commonly made of
   A. tempered glass.
   B. laminated glass.
   C. plate glass.
   D. glazed glass.

_____ 8. The most common locks on the market today are
   A. mortise locks.
   B. cylindrical locks.
   C. padlocks.
   D. rim locks.

9. If you can see the hardware of a door, it is a(n)
   A. sliding door.
   B. inward-swinging door.
   C. outward-swinging door.
   D. overhead door.

_____ 10. The part of the door lock that catches and holds the door frame is called the
   A. latch.
   B. operator lever.
   C. deadbolt.
   D. lock body.

_____ 11. Which types of tools are often used to force entry into buildings?
   A. Striking tools
   B. Cutting tools
   C. Prying tools
   D. Through-the-lock tools

_____ 12. Company officers usually select both the point of entry and the
   A. rate of entry.
   B. equipment to be used.
   C. method to be used.
   D. point of exit.

_____ 13. The slab door typically used for entrance doors because it is heavy and difficult to force open is a
   A. metal door.
   B. solid-core door.
   C. hollow-core door.
   D. glass door.

_____ 14. Another term for the doorknob is the
   A. latch.
   B. operator lever.
   C. deadbolt.
   D. lock body.

_____ 15. Which types of tools are used to generate a force directly on an object or another tool?
   A. Striking tools
   B. Cutting tools
   C. Prying tools
   D. Through-the-lock tools

_____ 16. Larger pieces or panes of glass are called
   A. annealed glass.
   B. tempered glass.
   C. laminated glass.
   D. plate glass.

_____ 17. The adz, the pick, and the claw are all incorporated into the
   A. hammer.
   B. Halligan tool.
   C. maul.
   D. pick axe.

_____ 18. The windows that are similar to sliding doors are called
   A. jalousie windows.
   B. casement windows.
   C. horizontal-sliding windows.
   D. projected windows.

_____ 19. Doors that have a wood frame inset with solid wood panels are called
   A. tempered doors.
   B. ledge doors.
   C. panel doors.
   D. slab doors.

_____ 20. Which type of circular saw blade can be used to cut through metal doors, locks, and gates?
   A. Steel blade
   B. Masonry-cutting blade
   C. Carbide-tipped blade
   D. Metal-cutting blade

_____ 21. A tool that has evolved from the use of a large log and is used to force doors or breach walls is a
   A. hammer.
   B. maul.
   C. lift.
   D. battering ram.

Labeling

Label the following diagram with the correct terms.

1. Basic parts of a door lock.

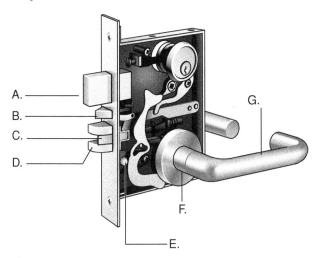

Figure 12-22

A. ______________________

B. ______________________

C. ______________________

D. ______________________

E. ______________________

F. ______________________

G. ______________________

2. Parts of mortise locks.

Figure 12-26

Courtesy of IR Safety and Security - Americas.

A. ______________________

B. ______________________

C. ______________________

D. ______________________

E. ______________________

F. ______________________

G. ______________________

H. ______________________

I. ______________________

J. ______________________

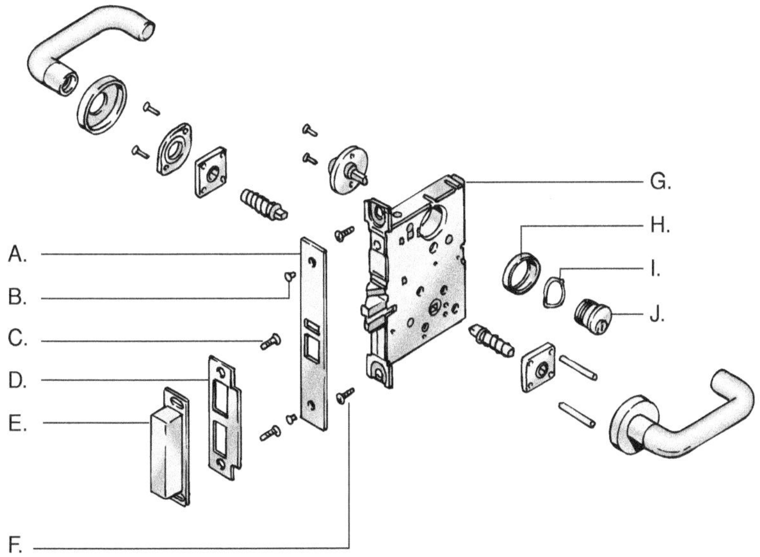

Vocabulary

Define the following terms using the space provided.

1. Casement windows:

2. Projected windows:

3. Rabbet:

4. Mortise locks:

5. Tempered glass:

6. K tool:

7. Cylindrical locks:

8. Jalousie windows:

_____

_____

_____

9. Jamb:

_____

_____

_____

Fill-in
Read each item carefully, then complete the statement by filling in the missing word(s).

1. _____ _____ are used to cut metal components such as bolts, padlocks, and chains.

2. A(n) _____ _____ is a small hydraulic spreader operated by a hand-powered pump.

3. Forcible entry is usually required at emergency incidents where time is a(n) _____ factor.

4. _____-_____ windows have two movable sashes that move freely up and down.

5. The quickest way to force entry through a(n) _____ roll-up door is to cut the door with a saw or torch.

6. _____ windows are similar in operation to jalousie windows, except that they usually have one large or two medium-sized glass panels instead of many small ones.

7. All tools should be kept in a(n) _____ state.

8. _____ provide air flow and light to the inside of the buildings, but can also provide emergency entrances.

9. _____-powered tools are portable and can be placed into operation quickly, but have limited power and operating times.

10. _____ doors are usually made of four glass panels with metal frames.

11. The circular saw blade that can stay sharp for long periods of time and can cut through hard surfaces or wood is the _____-tipped blade.

True/False
If you believe the statement to be more true than false, write the letter "T" in the space provided. If you believe the statement to be more false than true, write the letter "F."

_____ 1. The best point from which to attempt forcible entry to a structure is the door or window.

_____ 2. Interior walls are usually constructed of wood or metal studs and covered by plaster, gypsum, or sheetrock.

_____ 3. Fire fighters must consider, when making forcible entry, the need to secure the premises after operations are completed because they must never leave the premises in a state that would allow unauthorized entry.

_____ 4. A rabbet tool is a small hydraulic spreader operated by a hand-powered pump.

_____ 5. A pry axe should be used only for cutting.

_____ 6. The two most popular floor materials found in residences and commercial buildings are tile and steel.

_____ 7. Outward-opening doors are most often used in residential occupancies.

_____ 8. Double-pane glass is being used in many homes because it improves home insulation.

_____ 9. Duck-billed lock breakers are cutting tools used to snip off the shackles of a lock.

_____ 10. When breaking a window, always stand downwind.

_____ 11. The locking mechanisms on sliding doors are not strong and can be pried open.

Short Answer

Complete this section with short written answers using the space provided.

1. Identify the four general carrying tips that apply to all tools.

2. List the four categories of forcible entry tools.

3. Identify the four general safety tips for using tools.

4. List the four basic components of a door.

120 Fundamentals of Fire Fighter Skills

Word Fun

The following crossword puzzle is an activity provided to reinforce correct spelling and understanding of terminology associated with firefighting. Use the clues provided to complete the puzzle. Do not include spaces or punctuation when filling in the puzzle.

CLUES

Across

2. The pointed end of a pick axe, which can be used to make a hole or bite in a door, floor, or wall.
4. The prying part of the Halligan tool.
6. A type of door frame in which the stop for the door is cut into the frame.
9. A spring-loaded latch bolt or a gravity-operated steel bar that, after release by physical action, returns to its operating position and automatically engages the strike plate when it is returned to the closed position. (NFPA 80)
13. A cutting tool with a pry bar built into the cutting part of the tool.
14. The part of a doorway that secures the door to the studs in a building.
15. A small opening made to enable better tool access in forcible entry.
17. An entryway; the primary choice for forcing entry into a vehicle or structure.
18. A nonstructural interior wall that spans horizontally or vertically from support to support. The supports may be the basic building frame, subsidiary structural members, or other portions of the system. [ASCE/SEI 7:11.2] (NFPA 5000)
19. The most common types of locks on the market today, built to provide regular-duty or heavy-duty service. Several types of locking mechanisms are available, including keyways, combination wheels, and combination dials.

Down

1. A specially designed hand axe that serves multiple purposes. Similar to a Halligan bar, it can be used to pry, cut, and force doors, windows, and many other types of objects. Also called a multipurpose axe.
2. A type of glass that has additional strength so it can be formed in larger sheets but will still shatter upon impact.
3. A tool that is designed to fit between double doors equipped with panic bars.
5. A tool used to cut through thick metal objects, such as bolts, locks, and wire fences.
7. A tool that is used to remove lock cylinders from structural doors so the locking mechanism can be unlocked.
8. Surface- or interior-mounted lock on or in a door with a bolt that provides additional security.
10. The parts of a door or window that enable it to be locked or opened.
11. Glass or transparent or translucent plastic sheet used in windows, doors, skylights, or curtain walls. [ASCE/SEI 7:6.2] (NFPA 5000)
12. The forked end of a tool.
16. A combination tool, normally consisting of the Halligan tool plus a flat-head axe.

Fire Alarms

The following real case scenarios will give you an opportunity to explore the concerns associated with forcible entry. Read each scenario, and then answer each question in detail.

1. It is 3:00 in the afternoon when your ladder truck company is dispatched to a commercial structure fire in a food storage warehouse. Upon arrival, your ladder truck is assigned to force entry on side C of the building. Your Lieutenant performs a size-up and determines that the best route of entry will be through an overhead-rolling door. She tells you and your partner to gather the appropriate tools and force entry through the door. How do you proceed?

   _____
   _____
   _____
   _____
   _____
   _____

2. It is 1:30 in the morning when your engine company is dispatched to an alarm activation at an elementary school. Upon arrival, you find light smoke in the hallway and the alarm annunciator panel tells you that water is flowing. Upon his walk-around, your Captain finds a single sprinkler head that has activated in a classroom and reports that the fire has been contained. He directs your crew to force entry through the classroom doors and overhaul the fire. Upon size-up, you determine the entrance doors are glass, outward-opening doors with a steel frame. How should you proceed?

   _____
   _____
   _____
   _____
   _____
   _____

Fire Fighter II in Action

The following scenario will give you an opportunity to apply your firefighting knowledge and your fire department SOGs to the new information you learned while studying this chapter. Research your department's SOGs and answer the assignment in detail. Compare your answers with your classmates' and discuss similarities and obvious differences between your answers.

Your Lieutenant has given you the task of giving forcible entry training to your station. He has requested you focus on tools specifically, choosing the proper tool for the job at hand. This may have something to do with the axe handle you broke at the last fire…

1. Identify the four categories of forcible entry tools. Demonstrate proper use of each tool from these four categories that your company carries.

   _____
   _____
   _____
   _____
   _____

Skill Drills

Skill Drill 12-1: Forcing Entry into an Inward-Opening Door
Test your knowledge of the skill drill by filling in the correct words in the photo captions.

© Jones & Bartlett Learning. Photographed by Glen E. Ellman.

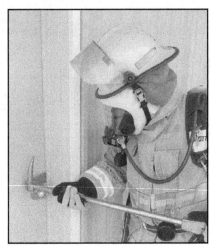
© Jones & Bartlett Learning. Photographed by Glen E. Ellman.

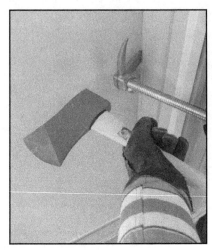
© Jones & Bartlett Learning. Photographed by Glen E. Ellman.

1. Size up the door, looking for any _____ _____. Inspect the door for the location and number of locks and mechanisms.

2. Place the _____ of the Halligan tool into the door frame between the door jamb and the door stop, near the lock, with the _____ end of the tool against the door.

3. Once the Halligan tool is in position, have your partner, on your command, drive the tool farther into the gap between the _____ _____ or stop and the door. Make sure that the tool is not driven into the door jamb itself.

© Jones & Bartlett Learning. Photographed by Glen E. Ellman.

4. Once the tool is past the stop but between the door and the jamb, push the Halligan tool toward the door to force it open. If more leverage is needed, your partner can slide the _____ _____ between the _____ of the Halligan tool and the door.

Skill Drill 12-4: Forcing Entry Through a Wooden Double-Hung Window

Test your knowledge of the skill drill by filling in the correct words in the photo captions.

© Jones & Bartlett Learning. Photographed by Glen E. Ellman.

1. Size up the window for any safety hazards and locate the _____ _____.

© Jones & Bartlett Learning. Photographed by Glen E. Ellman.

2. Place the _____ _____ of the Halligan tool under the bottom sash in line with the locking mechanism.

© Jones & Bartlett Learning. Photographed by Glen E. Ellman.

3. Pry the _____ _____ upward to displace the locking mechanism. Secure the window so that it does not close.

Ladders

Workbook Activities

The following activities have been designed to help you. Your instructor may require you to complete some or all of these activities as a regular part of your fire fighter training program. You are encouraged to complete any activity that your instructor does not assign as a way to enhance your learning in the classroom.

Chapter Review

The following exercises provide an opportunity to refresh your knowledge of this chapter.

Matching

Match each of the terms in the left column to the appropriate definition in the right column.

_____ 1. Stop   **A.** A piece of material that prevents the fly section(s) of a ladder from overextending and collapsing the ladder

_____ 2. Rung   **B.** A ladder crosspiece that provides a climbing step

_____ 3. Halyard   **C.** The end of the ladder that is placed against the ground when the ladder is raised

_____ 4. Staypoles   **D.** The very top of the ladder

_____ 5. Dogs   **E.** Part of a Bangor ladder

_____ 6. Tip   **F.** The ladder component that supports the rungs

_____ 7. Beam   **G.** The top and bottom surfaces of a trussed ladder

_____ 8. Egress   **H.** The rope or cable used to extend or hoist the fly section(s) of an extension ladder

_____ 9. Rail   **I.** A method of exiting from an area or a building

_____ 10. Butt   **J.** A mechanical locking device used to secure the fly section(s) of a ladder

Multiple Choice

Read each item carefully, and then select the best response.

_____ 1. The rope or cable used to extend the fly section is the
     A. dog.
     B. guide.
     C. halyard.
     D. lift.

CHAPTER 13

_____ 2. Ladders that are designed to allow access to attic scuttle holes and confined areas are
   A. folding ladders.
   B. pompier ladders.
   C. scaling ladders.
   D. combination ladders.

_____ 3. When a fire fighter stands between the ladder and the structure, grasps the beams, and leans to pull the ladder into the structure, he or she is
   A. butting the ladder.
   B. guiding the ladder.
   C. heeling the ladder.
   D. dogging the ladder.

_____ 4. The most important safety check is confirming the
   A. proximity to direct flames.
   B. location of overhead utility lines.
   C. identification of stable, level surfaces.
   D. size of the structure.

_____ 5. Establishing verbal contact as quickly as possible is important when rescuing
   A. an unconscious patient.
   B. an infant.
   C. an elderly person.
   D. any person.

_____ 6. A ladder that has no halyard, is generally short, and is designed for attic access is the
   A. pompier ladder.
   B. Bangor ladder.
   C. Fresno ladder.
   D. roof ladder.

_____ 7. Ladder gins are used to access
   A. below-grade sites.
   B. above-grade sites.
   C. grade-level positions.
   D. interior above-grade sites.

_____ 8. When climbing a ladder, the fire fighter's eyes should be looking
   A. forward, with an occasional glance upward.
   B. upward.
   C. downward.
   D. at the ladder's tip.

_____ 9. The _____ is a small grooved wheel that is used to change the direction of the halyard pull.
   A. pulley
   B. foot
   C. heel
   D. stop

_____ 10. Transferring the weight of the user to the beams is done through the
   A. halyards.
   B. rungs.
   C. butt spurs.
   D. tie rods.

_____ 11. Staypoles are required on ladders of
  A. 10 to 20 feet (3 to 6 meters).
  B. 20 to 30 feet (6 to 9 meters).
  C. 30 to 40 feet (9 to 12 meters).
  D. 40 feet (12 meters) or greater.

_____ 12. Most ladders are carried with the
  A. tip forward.
  B. fly-section forward.
  C. butt end forward.
  D. beam on top.

_____ 13. The horizontal-bending test evaluates the
  A. manufacturer's specifications.
  B. structural strength of the ladder.
  C. service testing of the ladder.
  D. extension hardware on the ladder.

_____ 14. The two common techniques for raising portable ladders are the
  A. team and self raises.
  B. beam and rung raises.
  C. beam and halyard raises.
  D. team and aerial raises.

_____ 15. Butt spurs prevent the ladder from
  A. losing contact with the exterior of the structure.
  B. damaging the structure.
  C. chaffing other surfaces.
  D. slipping out of position.

_____ 16. Ladders with staypoles or tormentors are typically referred to as
  A. aerial ladders.
  B. portable ladders.
  C. Bangor ladders.
  D. Fresno ladders.

_____ 17. What is the common rule of thumb when identifying the length of ladder to use on a structure?
  A. The tip of the ladder is in contact with the structure.
  B. The butt of the ladder is in contact with the structure.
  C. At least five ladder rungs show above the roofline.
  D. The ladder provides access to all possible incident outcomes.

_____ 18. The number of fire fighters required to raise a ladder depends on the
  A. length and width of the ladder.
  B. weight and target surface of the ladder.
  C. width and clearance of the ladder.
  D. length and weight of the ladder.

_____ 19. The proper climbing angle for maximum load capacity and strength is
  A. 30 degrees.
  B. 45 degrees.
  C. 60 degrees.
  D. 75 degrees.

_____ 20. A straight ladder equipped with retractable hooks to secure the tip of the ladder to a pitched roof is a
  A. roof ladder.
  B. Bangor ladder.
  C. combination ladder.
  D. Fresno ladder.

_____ 21. In most cases, a single fire fighter can safely carry a straight or roof ladder
  A. less than 18 feet (5.5 meters) long.
  B. less than 24 feet (7 meters) long.
  C. less than 4 feet (1.2 meters) wide.
  D. less than 2 feet (0.6 meters) wide.

_____ 22. Most portable ladders are designed to support a weight of
   A. 500 pounds (226.8 kilograms).
   B. 750 pounds (340.2 kilograms).
   C. 1000 pounds (453.6 kilograms).
   D. 1500 pounds (680.4 kilograms).

_____ 23. The rail is the
   A. top section of a solid beam.
   B. top section of a trussed beam.
   C. top or bottom section of a trussed beam.
   D. handrail at the tip of the ladder.

_____ 24. Pawls, ladder locks, or rung locks are also referred to as
   A. roof hooks.
   B. dogs.
   C. guides.
   D. truss locks.

_____ 25. Fire service portable ladders are limited to a maximum length of
   A. 25 feet (7.6 meters).
   B. 50 feet (15.2 meters).
   C. 75 feet (22.9 meters).
   D. 100 feet (30.5 meters).

_____ 26. The part of an extension ladder that is raised or extended from the bed section is the
   A. fly section.
   B. elevating section.
   C. lift section.
   D. aerial section.

_____ 27. The metal bar that runs from one beam of the ladder to the other and keeps the beams from separating is the
   A. butt spur.
   B. rung.
   C. rail.
   D. tie rod.

_____ 28. A truss block is a piece that connects the two
   A. rails of a trussed beam.
   B. rungs of a trussed beam.
   C. pulleys of an I-beam.
   D. rungs of an I-beam.

128 FUNDAMENTALS OF FIRE FIGHTER SKILLS

Labeling
Label the following diagram with the correct terms.

1. Basic components of a straight ladder.

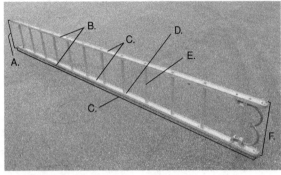

A. _____
B. _____
C. _____
D. _____
E. _____
F. _____

© Jones & Bartlett Learning. Photographed by Glen E. Ellman.

Figure 13-2

Vocabulary
Define the following terms using the space provided.

1. Guides:

2. Ladder belt:

3. Heat sensor label:

4. Tie rod:

5. Roof hooks:

6. Protection plates:

7. Pawls:

8. Bed section:

9. Halyard:

10. Pulley:

Fill-in
Read each item carefully, and then complete the statement by filling in the missing word(s).

1. A(n) _____ ladder is an assembly of two or more ladder sections that can be extended or retracted to adjust the length.

2. Fire fighters who are working from a ladder should use a ladder belt or a(n) _____ _____ to secure themselves to the ladder.

3. A(n) _____ ladder can be converted from a straight ladder to a stepladder configuration.

4. When dismounting a ladder, the fire fighter should try to maintain contact with the ladder at _____ points.

5. The _____ ladder is one of the most functional, versatile, and rapidly deployable tools used by fire fighters.

6. A rope, a rope-hose tool, or _____ can be used to secure a ladder in place.

7. A(n) _____ ladder is a single-section, fixed-length ladder.

8. The _____ serve as the steps of a ladder.

9. _____ ladders are permanently mounted, power-operated ladders with a working length of at least 50 feet (15.2 meters).

10. Ladders provide elevated platforms for _____ as well as for fire fighters.

11. Ladders should always be inspected and maintained in accordance with the _____ recommendations.

True/False
If you believe the statement to be more true than false, write the letter "T" in the space provided. If you believe the statement to be more false than true, write the letter "F."

_____ 1. Ladders consist of two rungs connected by a series of parallel beams.

_____ 2. The butt and the heel of a ladder are at opposite ends of a portable ladder.

_____ 3. Pulling on the halyard extends the bed sections of a combination ladder.

_____ 4. The ladder is one of the fire fighter's basic tools.

_____ 5. In general, ladder manufacturers recommend that the fly sections be placed toward the structure.

_____ 6. A fire fighter working from a ladder is in a less stable position than a fire fighter working on the ground.

_____ 7. During a three-fire-fighter shoulder carry, the middle fire fighter should be on the opposite side of the other two.

_____ 8. When roof ladders are properly attached, they will not support the weight of the ladder and a fire fighter.

_____ 9. When using an extension ladder, never wrap the halyard around your hand.

_____ 10. A ladder can be used as a work platform.

_____ 11. Self-contained breathing apparatus (SCBA) is not required when working on the roof at a chimney fire.

_____ 12. There are two tips on a portable ladder.

Short Answer

Complete this section with short written answers using the space provided.

1. In the proper order, list the five steps to apply a leg lock to work from a ladder.

2. Identify five basic safety concerns when using ladders.

3. Describe the three basic types of beam construction.

4. Identify five fundamental ladder maintenance tasks.

Word Fun

The following crossword puzzle is an activity provided to reinforce correct spelling and understanding of terminology associated with firefighting. Use the clues provided to complete the puzzle. Do not include spaces or punctuation when filling in the puzzle.

Clues

Across

1. A piece of material that prevents the fly sections of a ladder from becoming overextended, leading to collapse of the ladder.
3. A single ladder equipped with hooks at the top end of the ladder. (NFPA 1931)
5. The top or bottom piece of a trussed beam assembly used in the construction of a trussed ladder. Also, the top and bottom surfaces of an I-beam ladder.
6. A compliant equipment item that is intended for use as a positioning device for a person on a ladder.
9. Rope used on extension ladders for the purpose of raising a fly section(s).
10. Component of a ground ladder support that remains in contact with the lower support surface to reduce slippage.
12. To go or come out; to exit from an area or a building.
14. The lowest or widest section of an extension ladder.
16. Poles attached to each beam of the base section of extension ladders, which assist in raising the ladder and help provide stability of the raised ladder.

Down

2. A device with a free-turning, grooved metal wheel (sheave) used to reduce rope friction. Side plates are available for a carabiner to be attached.
4. The spring-loaded, retractable, curved metal pieces that allow the tip of a roof ladder to be secured to the peak of a pitched roof. They fold outward from each beam at the top of a roof ladder.
5. The ladder crosspieces, on which a person steps while ascending or descending.
7. The main structural side of a ground ladder.
8. An A-shaped structure formed with two ladder sections. It can be used as a makeshift lift when raising a trapped person. One form of the device is called an A-frame hoist.
10. The end of the beam that is placed on the ground, or other lower support surface, when ground ladders are in the raised position.
11. Devices attached to a fly section(s) to engage ladder rungs near the beams of the section below for the purpose of anchoring the fly section(s).
13. A measurement of the angle used in road design and expressed as a percentage of elevation change over distance.
15. The very top of the ladder.

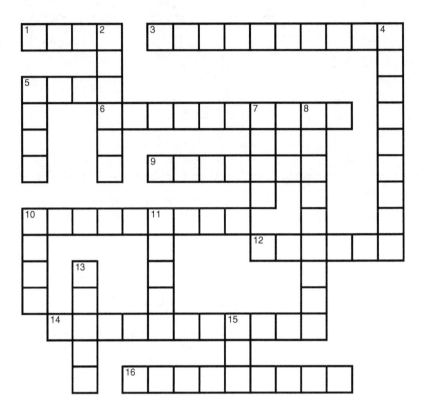

Fire Alarms

The following real case scenarios will give you an opportunity to explore the concerns associated with ladders. Read each scenario, and then answer each question in detail.

1. It is 10:45 in the morning when your company is dispatched to an older two-story apartment complex. The fire is located in two units on the second floor. Upon arrival, the incident commander assigns you to Division C. Your Captain receives instructions to deploy a roof ladder utilizing the extension ladder already in place on side C of the structure. The Captain tells you and your partner to remove the roof ladder from the apparatus and place it on the roof. How should you proceed?

   _____
   _____
   _____
   _____
   _____
   _____

2. It is 11:30 in the evening when your engine company is dispatched to a two-story residential building for a structure fire with a life threat. As you exit your engine, you can hear cries for help. When you look toward the structure, you see a middle-aged woman straddling a second-floor window with smoke pouring out from behind her. The adjacent window is engulfed with fire. Your Lieutenant tells you to ladder the window for a rescue. How should you proceed?

   _____
   _____
   _____
   _____
   _____
   _____

Fire Fighter II in Action

The following scenario will give you an opportunity to apply your firefighting knowledge and your fire department SOGs to the new information you learned while studying this chapter. Research your department's SOGs and answer the assignment in detail. Compare your answers with your classmates' and discuss similarities and obvious differences between your answers.

You have been ordered to place a roof ladder in service then vent a roof of a single-family dwelling. The roof is pitched, with the gutter line being 24' (7.3 m) above ground level.

1. What ladders will you select?

   _____
   _____
   _____
   _____
   _____

Skill Drills

Skill Drill 13-2: One-Fire-Fighter Carry
Test your knowledge of the skill drill by filling in the correct words in the photo captions.

© Jones & Bartlett Learning. Photographed by Glen E. Ellman.

1. Start with the ladder mounted in a bracket or standing on one _____. Locate the _____ of the ladder. Place an arm between two rungs of the ladder just to one side of the _____ rung.

© Jones & Bartlett Learning. Photographed by Glen E. Ellman.

2. Lift the top _____ of the ladder and rest it on your _____.

© Jones & Bartlett Learning. Photographed by Glen E. Ellman.

3. Walk carefully with the _____ end first.

Skill Drill 13-3: Two-Fire-Fighter Shoulder Carry

Test your knowledge of this skill drill by placing the photos below in the correct order. Number the first step with a "1," the second step with a "2," and so on.

_____ The butt spurs are covered with a gloved hand while the ladder is transported.

_____ Both fire fighters place one arm between two rungs and, on the leader's command, lift the ladder onto their shoulders. The ladder is carried butt end first.

_____ Start with the ladder mounted in a bracket or standing on one beam. Both fire fighters are positioned on the same side of the ladder. Facing the butt end of the ladder, one fire fighter is positioned near the butt end of the ladder and a second fire fighter is positioned near the tip of the ladder.

Skill Drill 13-13: Tying the Halyard

Test your knowledge of the skill drill by filling in the correct words in the photo captions.

1. _____ the excess halyard rope around two rungs of the ladder and pull the rope tight across the _____ of the two rungs.

2. Tie a(n) _____ _____ around the upper rung and the vertical section of the halyard.

3. _____ the clove hitch tight.

4. Place an overhand _____ knot as close to the clove hitch as possible to prevent _____.

Skill Drill 13-19: Use a Leg Lock to Work from a Ladder

Test your knowledge of this skill drill by placing the photos below in the correct order. Number the first step with a "1," the second step with a "2," and so on.

_____ Secure your foot against the next lower rung or the beam of the ladder. Use your thigh for support and step down one rung with the opposite foot.

_____ Once your leg is between the rungs, bend your knee and bring your foot back under the rung and through to the climbing side of the ladder.

_____ Climb to the desired work height and step up to one more rung.

_____ The use of a leg lock enables you to have two hands free for a variety of tasks.

_____ Note the side of the ladder where the work will be performed. Extend your leg between the rungs on the side opposite the side you will be working.

Search and Rescue

Workbook Activities

The following activities have been designed to help you. Your instructor may require you to complete some or all of these activities as a regular part of your fire fighter training program. You are encouraged to complete any activity that your instructor does not assign as a way to enhance your learning in the classroom.

Chapter Review

The following exercises provide an opportunity to refresh your knowledge of this chapter.

Matching
Match each of the terms in the left column to the appropriate definition in the right column.

_____ 1. Secondary search
_____ 2. Primary search
_____ 3. Search and rescue
_____ 4. Search
_____ 5. Rescue techniques
_____ 6. Transitional attack
_____ 7. Rescue
_____ 8. Rekindle
_____ 9. Thermal imager
_____ 10. Search rope

A. A return to flaming combustion after apparent but incomplete extinguishment
B. An offensive fire attack initiated by an exterior, indirect handline operation into the fire compartment to initiate cooling while transitioning into an interior attack.
C. The process of looking for living victims who are in danger
D. Assists, drags, and carries
E. Similar to a video camera except that it captures heat display images instead of visible light images
F. The removal of a person from danger
G. A more thorough search undertaken after the fire is under control
H. May be performed by any fire company or unit
I. Quick initial search for victims
J. A guide rope that allows fire fighters to maintain contact with a fixed point

Multiple Choice
Read each item carefully, and then select the best response.

_____ 1. Search ropes should be used to
   A. help fire fighters maintain their orientation.
   B. keep search teams connected.
   C. search wide open spaces.
   D. both A and C.

_____ 2. If a victim is capable of walking, rescuers may only need to use the
   A. one-person walking assist.
   B. two-person seat carry.
   C. cradle-in-arms carry.
   D. three-person walking assist.

CHAPTER 14

_____ 3. To rescue a victim through a window, raise the ladder and secure it in the rescue position with the tip
   A. above the windowsill.
   B. in the open window.
   C. just below the windowsill.
   D. upwind from the window.

_____ 4. When rescuing a heavy adult using a ladder, the rescuer should
   A. get more help—three rescuers at a minimum.
   B. use two ladders.
   C. place two ladders side by side.
   D. all of the above.

_____ 5. A webbing sling provides a secure grip around
   A. the victim's upper body.
   B. the victim's waist.
   C. the victim's arms.
   D. the victim's legs and feet.

_____ 6. When rescuing an unconcious victim on a ladder, the fire fighter on the ladder should
   A. maintain continuous eye contact with the victim.
   B. maintain eye contact with the second fire fighter.
   C. face the victim.
   D. remain on the ladder until a rescue team can assist the descent.

_____ 7. After the fire is under control, fire fighters should begin a
   A. primary search.
   B. secondary search.
   C. rescue.
   D. safety search.

_____ 8. When a ladder rescue involves a conscious victim, the fire fighter should
   A. establish verbal contact.
   B. urge the victim to jump.
   C. have the victim climb down facing the fire fighter.
   D. have the victim exit head first.

_____ 9. After the area immediately around the fire is searched in an apartment building, the next priority is to search the
   A. area directly above the fire.
   B. area directly below the fire.
   C. highest floors in the building.
   D. hallways and exits.

_____ 10. A search begins with the areas in which
   A. the greatest number of hazards exist.
   B. the building experiences the greatest traffic.
   C. occupants are expected.
   D. victims are at the greatest risk.

_____ 11. Long backboard rescues are used to
   A. carry a conscious victim down a ladder.
   B. remove a victim from a vehicle.
   C. aid with a one-rescuer drag.
   D. aid with a fire fighter drag.

_____ 12. Assisting someone down a ladder carries a considerable risk to
   A. the victim.
   B. the fire fighters.
   C. the safety officer.
   D. both A and B.

_____ 13. Status and results from search operations needs to be communicated to the
   A. secondary search team.
   B. incident commander.
   C. safety officer.
   D. rapid intervention company/crew.

_____ 14. When conducting searches in high-rise buildings, it is important to work
   A. from the bottom floor up.
   B. from the middle floors out.
   C. from the walls to the middle of the rooms.
   D. as teams, coordinating searches.

_____ 15. When a building is occupied, fire fighters should first rescue the occupants who are
   A. in the most immediate danger.
   B. in the least danger.
   C. the most easily accessed.
   D. closest to the exits.

_____ 16. During an oriented-vent-enter-isolate-search, firefighters
   A. work in pairs.
   B. carry no radios.
   C. enter the fire room at the same time.
   D. keep the hallway door open for egress.

_____ 17. The three most important senses during a search are
   A. sight, sound, and taste.
   B. touch, sight, and taste.
   C. sight, sound, and touch.
   D. sound, taste, and touch.

_____ 18. The two-person seat carry is used when the victim
   A. is very large.
   B. must be carried up or down stairs.
   C. is a child.
   D. is disabled or paralyzed.

_____ 19. The fire fighter drag utilizes the victim's
   A. clothing as a handle.
   B. tied wrists.
   C. weight to assist the movement.
   D. ability to assist moving.

_____ 20. The clothes drag is used to move a victim who is on the floor and
   A. is too heavy for one rescuer.
   B. must be carried up or down stairs.
   C. is disabled or paralyzed.
   D. is difficult to reach.

_____ 21. A method of searching is:
   A. standard search.
   B. oriented search.
   C. O-VEIS.
   D. A, B, and C.

_____ 22. The oriented search method is well suited to searching:
   A. High rise buildings.
   B. Warehouses.
   C. Residential settings.
   D. Schools.

Vocabulary

Define the following terms using the space provided.

1. Exit assist:

2. Shelter-in-place:

3. Two-in/two-out rule:

4. Primary search:

5. Transitional attack:

Fundamentals of Fire Fighter Skills

Fill-in
Read each item carefully, and then complete the statement by filling in the missing word(s).

1. Saving _____ has the highest priority at a fire scene.
2. Search team members may have to _____ to stay below layers of hot gases and smoke.
3. Occupants who are asleep are at a(n) _____ risk than occupants who are awake.
4. The incident commander is responsible for managing the level of _____ during emergency operations.
5. Search teams must have a(n) _____ escape route in case fire conditions change.
6. Searched rooms should be _____ so other searchers will know they have been searched.
7. The decision to enter a burning building to search for living victims is one that may be made by the _____ _____, the _____ _____ _____ or a _____ _____.
8. The application of a solid stream into the room of origin, resulting in dramatic cooling, is called _____ _____ _____.
9. Upon completion of all searches, the priority can shift to _____ and _____ the fire.
10. _____ _____ to determine whether they are stable is a dangerous practice and should be discouraged.
11. Searchers need to be aware of the _____ _____ _____ in a building and avoid actions that could place them in a dangerous position.
12. The _____ _____ is the most commonly taught search method used by fire fighters.

True/False
If you believe the statement to be more true than false, write the letter "T" in the space provided. If you believe the statement to be more false than true, write the letter "F."

_____ 1. More fire fighters will be needed for search and rescue operations in a nursing home than in a similar-sized office building.

_____ 2. It is justifiable to risk the safety of fire fighters if there is a potential to save lives.

_____ 3. In some situations, the best option is to shelter occupants in place.

_____ 4. It is often difficult to obtain accurate information from people who have just escaped from a burning building.

_____ 5. Most people who realize they are in danger will attempt to escape on their own.

_____ 6. The overall plan for a fire incident must focus on the life-safety priority as long as search and rescue operations are still under way.

_____ 7. Search and rescue operations should utilize only one search and rescue team.

_____ 8. Observations of the size and arrangement of the building can provide valuable information when trying to determine a search plan.

_____ 9. The two-in/two-out rule can be broken if there is an imminent life-threatening situation.

_____ 10. The standard search is conducted with a search team of two or more members.

Short Answer
Complete this section with short written answers using the space provided.

1. Describe three benefits of thermal imagers.

2. List six pieces of search and rescue equipment.

3. Identify the six guidelines fire fighters need to remember during search and rescue operations.

4. Identify five pieces of valuable information a preincident plan can provide for search and rescue operations.

5. Identify the four simple carries that can be used to move a victim who is conscious and responsive, but incapable of standing or walking.

6. Identify four considerations for search and rescue size-up.

7. Explain why sounding a floor is considered dangerous and should be discouraged.

8. Describe two disadvantages of the standard search:

9. Explain when O-VEIS should be used, and when it should not be attempted.

Fire Alarms

The following real case scenarios will give you an opportunity to explore the concerns associated with search and rescue. Read each scenario, and then answer each question in detail.

1. You are the leader of a search and rescue team about to enter a carpet warehouse. Two employees are missing and thought to be located in the fire building. The warehouse has sprinklers, and the fire is confined. Your officer warns that the inside floor plan is complex owing to the numerous remodels. To make matters worse, the smoke is thick and black because of the burning carpet, and it is cold and hugging the floor because of the activation of the sprinkler system. Knowing that this is a very dangerous task, you want to ensure the safety of yourself and your crew. How should you proceed?

2. It is 1:00 in the morning when you are dispatched to an old three-story apartment complex. The main stairway is located in the center of the building. The fire is located on the first floor. A suppression crew has been sent to the seat of the fire. There are mixed reports on whether the building has been evacuated. Your engine company has been assigned to the third floor for search and rescue. How will you proceed?

Fundamentals of Fire Fighter Skills

Fire Fighter II in Action

The following scenario will give you an opportunity to apply your firefighting knowledge and your fire department SOGs to the new information you learned while studying this chapter. Research your department's SOGs and answer the assignment in detail. Compare your answers with your classmates' and discuss similarities and obvious differences between your answers.

Your first and second response districts have numerous "big box" stores, as well as several warehouses and large industrial buildings. Your fire department has 30-minute, 4500-psi self-contained breathing apparatus (SCBA) air bottles which limit your ability to search these large structures.

1. What is your fire department SOG on large area search and air management?

   _____
   _____
   _____
   _____
   _____
   _____

2. What other concerns do you have with searching for victims in these large buildings?

   _____
   _____
   _____
   _____
   _____
   _____

Skill Drills

Skill Drill 14-5: Performing a Two-Person Extremity Carry
Test your knowledge of the skill drill by filling in the correct words in the photo captions.

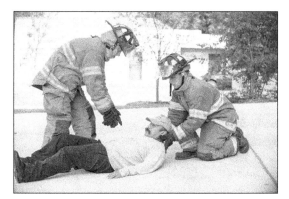

© Jones & Bartlett Learning. Photographed by Glen E. Ellman.

1. Two fire fighters help the victim to _____ up.

© Jones & Bartlett Learning. Photographed by Glen E. Ellman.

2. The first fire fighter kneels _____ the victim, reaches under the victim's arms, and grasps the victim's _____.

© Jones & Bartlett Learning. Photographed by Glen E. Ellman.

3. The second fire fighter backs in between the victim's _____, reaches around, and grasps the victim behind the _____.

© Jones & Bartlett Learning. Photographed by Glen E. Ellman.

4. The first fire fighter gives the command to stand and carry the victim away, walking straight ahead. Both fire fighters must _____ their movements.

Skill Drill 14-7: Performing a Two-Person Chair Carry

Test your knowledge of the skill drill by filling in the correct words in the photo captions.

© Jones & Bartlett Learning. Photographed by Glen E. Ellman.

1. One fire fighter stands behind the seated victim, reaches down, and grasps the _____ of the chair.

© Jones & Bartlett Learning. Photographed by Glen E. Ellman.

2. The fire fighter tilts the chair backward on its rear legs so that the second fire fighter can step back between the legs of the chair and grasp the tips of the chair's _____ _____. The victim's legs should be _____ the legs of the chair.

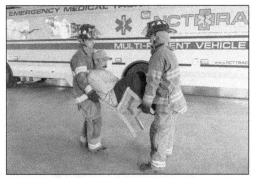

© Jones & Bartlett Learning. Photographed by Glen E. Ellman.

3. When both fire fighters are correctly positioned, the fire fighter _____ the chair gives the command to lift and walk away. Because the chair carry may force the victim's head _____, watch the victim for airway problems.

Skill Drill 14-17: Rescuing an Unconscious Victim from a Window

Test your knowledge of the skill drill by filling in the correct words in the photo captions.

1. Place the tip of the ladder just _____ the windowsill.

2. One fire fighter enters to rescue the victim. The second fire fighter _____ to the window.

3. The fire fighter waiting on the ladder places both hands on the rungs, with one leg straight and the other horizontal to the ground with the knee at an angle of _____ _____. The interior fire fighter passes the victim through the window and onto the ladder, keeping the victim's _____ toward the ladder.

4. The victim is lowered so that he or she _____ the fire fighter's leg. The fire fighter's arms should be positioned under the victim's arms, holding on to the _____. Step down one rung at a time, transferring the victim's weight from one leg to the other. The victim's arms can also be secured around the fire fighter's neck.

Ventilation

Workbook Activities

The following activities have been designed to help you. Your instructor may require you to complete some or all of these activities as a regular part of your fire fighter training program. You are encouraged to complete any activity that your instructor does not assign as a way to enhance your learning in the classroom.

Chapter Review

The following exercises provide an opportunity to refresh your knowledge of this chapter.

Matching

Match each of the terms in the left column to the appropriate definition in the right column.

_____ 1. Pitched roofs
_____ 2. Roof decking
_____ 3. Chase
_____ 4. Laths
_____ 5. Rafters
_____ 6. Flashover
_____ 7. Ejectors
_____ 8. Cockloft
_____ 9. Parapet walls
_____ 10. Transitional attack

A. The open space between the top-floor ceiling and the roof of a building
B. Can occur when the air in a room is superheated and exposed combustibles in the space are near their ignition point
C. Thin parallel strips of wood used to make the supporting structure for roof tiles
D. Applying water through a window or door from a safe location as close to the fire as possible just before or at the same time ventilation occurs
E. Electric or gasoline fans used in negative-pressure ventilation
F. The rigid roof component made of wooden boards, plywood sheets, or metal panels
G. Solid structural components that support a roof
H. Open space within walls for wires and pipes
I. Have a visible slope that provides for rain, ice, and snow runoff
J. Freestanding walls on a flat roof that extend above the roofline

Multiple Choice

Read each item carefully, and then select the best response.

_____ 1. Negative-pressure ventilation fans are called
  A. ventilators.
  B. conductors.
  C. smoke ejectors.
  D. HVAC systems.

CHAPTER 15

_____ 2. Which type of ventilation occurs when fans are used to force clean air into a space to displace smoke?
   A. Positive-pressure ventilation
   B. Natural ventilation
   C. Hydraulic ventilation
   D. Negative-pressure ventilation

_____ 3. A cut that works well on metal roofing because it prevents the decking from rolling away is a
   A. kerf cut.
   B. rectangular cut.
   C. louver cut.
   D. triangular cut.

_____ 4. Ventilation openings should never be
   A. opened directly into the atmosphere.
   B. between the fire fighters and the escape route.
   C. created without incident commander (IC) direction.
   D. opened before proper sounding.

_____ 5. A triangular examination hole created with three small cuts is a
   A. kerf cut.
   B. rectangular cut.
   C. louver cut.
   D. triangular cut.

_____ 6. Fire fighters should ask themselves what before opening a door, taking out a window, or putting a hole in a roof at a structure fire?
   A. Why am I ventilating?
   B. Where do I want to accomplish the ventilation?
   C. When do I want to perform the ventilation?
   D. A, B, and C

_____ 7. A cut that can create a large opening quickly and is particularly suitable for flat or sloping roofs with plywood decking is a
   A. kerf cut.
   B. rectangular cut.
   C. louver cut.
   D. triangular cut.

_____ 8. When fire fighters need quick or immediate ventilation, they often use
   A. natural ventilation.
   B. mechanical ventilation.
   C. hydraulic ventilation.
   D. vertical ventilation.

_____ 9. Trusses are connected with heavy-duty staples or by
   A. triangular plates.
   B. gusset plates.
   C. plate locks.
   D. truss plates.

_____ 10. The firefighting operation of removing smoke and heat from a structure by opening windows and doors or making holes in a roof is called
   A. ventilation.
   B. convection.
   C. conduction.
   D. smoke inversion.

_____ 11. All roofs have two major components:
   A. a support structure and a roof shoring.
   B. beams and rafters.
   C. a support structure and a roof covering.
   D. a platform and a support system.

_____ 12. A bearing wall is used
   A. as an exterior wall.
   B. to support the weight of a floor or roof.
   C. as an interior wall.
   D. to extend the firewall.

_____ 13. The act of closing the front door of a structure after forcible entry will
   A. create a strong downdraft.
   B. create a strong updraft.
   C. limit air supply and slow fire growth.
   D. create greater resistance to air movement.

_____ 14. A window or door failing or left open by mistake creates
   A. intentional ventilation.
   B. cross ventilation.
   C. accidental ventilation.
   D. stack effect.

_____ 15. Auto-exposure, fire jumping from floor to floor through an exterior window is also called
   A. vertical fire extension.
   B. traveling.
   C. exterior exposure.
   D. red sky.

_____ 16. Hydraulic ventilation is most useful for clearing smoke and heat out of a room because it creates
   A. a low-pressure area behind the nozzle.
   B. a high-pressure area behind the nozzle.
   C. a mist that traps smoke particles and heat.
   D. water vapor.

_____ 17. _____ fans are powered by internal combustion engines and can increase carbon monoxide levels if they run for significant periods of time after the fire is extinguished.
   A. Negative-pressure
   B. Horizontal
   C. Mechanical
   D. Positive-pressure

_____ 18. When breaking glass for ventilation purposes, the fire fighter should always use a(n)
   A. "all clear" call.
   B. hand tool.
   C. hose line.
   D. safety break before splintering.

_____ 19. Ventilating directly over the fire will
   A. produce the fastest impact on the fires behavior.
   B. exhaust the greatest amount of combustion products.
   C. cause stack effect.
   D. A and B

_____ 20. Horizontal ventilation is most effective when the opening goes directly
   A. to another space within the structure.
   B. into a stairwell.
   C. into the space where the fire is located.
   D. past the attack team.

_____ 21. The neutral plane is the interface at a vent between the hot gas flowing out of a compartment and
   A. water flowing into the compartment.
   B. cool air flowing into the compartment.
   C. hot smoke flowing out of the compartment.
   D. particulate flowing out of the compartment.

_____ 22. Which type of ventilation occurs when fans are used to pull smoke through openings?
   A. Positive-pressure ventilation
   B. Natural ventilation
   C. Hydraulic ventilation
   D. Negative-pressure ventilation

_____ 23. The easiest place to ventilate is usually
   A. the door.
   B. a skylight.
   C. a roof.
   D. windows.

_____ 24. A backdraft occurs when smoke, heat, and gases accumulate with a rich supply of partially burned fuels and are suddenly introduced to
   A. a flame.
   B. increased temperature.
   C. clean air.
   D. open fuels.

_____ 25. Which type of roof has a visible slope for rain or snow runoff?
   A. Bowstring roof
   B. Arched roof
   C. Flat roof
   D. Pitched roof

Fundamentals of Fire Fighter Skills

Vocabulary
Define the following terms using the space provided.

1. Smoke inversion:

2. Sounding:

3. Ventilation:

4. Ordinary construction:

5. Fire-resistive construction:

6. Gusset plates:

7. Vertical ventilation:

8. Stack effect:

9. Horizontal ventilation:

10. Flow path:

11. Fuel- limited fire:

Fundamentals of Fire Fighter Skills

Fill-in
Read each item carefully, and then complete the statement by filling in the missing word(s).

1. The greatest risk to fire fighters performing vertical ventilation is _____ _____.

2. _____ _____ ventilation creates a large opening ahead of the fire, removing a section of fuel for the fire to spread and increasing smoke and gas flow out of a building.

3. _____ ventilation is most useful after a fire is under control.

4. One means of controlling a _____ _____ fire is to limit the amount of oxygen that is available to the fire.

5. A _____ fire attack reduces the chance of a backdraft or flashover.

6. _____ ventilation takes advantage of the doors and windows on the same level as the fire, as well as any other horizontal openings that are available.

7. Fire fighters should be _____ from the ventilation openings so the wind will push the heat and smoke away.

8. The collapse of a(n) _____ truss roof is usually very sudden; for this reason, the presence of such a roof must be noted during preincident planning.

9. Positive-pressure fans operate at _____ velocity and can be very noisy.

10. _____ is a powerful force that can rapidly change the direction and speed of a fire and its flow path.

True/False
If you believe the statement to be more true than false, write the letter "T" in the space provided. If you believe the statement to be more false than true, write the letter "F."

_____ 1. Examination holes allow the team members to evaluate conditions under the roof and to verify the proper location for a ventilation opening.

_____ 2. Smoke can be cooled with automatic sprinkler systems.

_____ 3. HVAC or other building systems can cause fire to spread through leap-frogging.

_____ 4. A backdraft can occur when a building is charged with hot gases and oxygen is at the normal percentage in the building.

_____ 5. Vertical ventilation operations often involve opening or breaking a window.

_____ 6. Ventilation should occur as close to the fire as possible.

_____ 7. A strong concern that arises when structures have metal roofs is the release of flammable vapors, which can be the result of leaking roof coverings.

_____ 8. During ventilation, cutting several smaller holes is better than making one large hole.

_____ 9. Creating roof ventilation openings requires the use of many different tools and techniques, depending on the type of construction and roof decking.

_____ 10. When attacking basement fires, the use of interior stairs for ventilation should not be considered a safe option because of the danger of firefighters being in the exhaust flow path of hot gases from the fire.

Short Answer

Complete this section with short written answers using the space provided.

1. Describe the impact door control can have on fire growth.

2. Describe the difference between a primary cut and a secondary cut.

3. What is the objective of any roof ventilation operation?

4. List five indicators that it is time for immediate retreat from the roof of a structure.

5. List the three tactical priorities in structural firefighting operations and how the tactical priorities affect ventilation operations.

Fundamentals of Fire Fighter Skills

Fire Alarms

The following real case scenarios will give you an opportunity to explore the concerns associated with ventilation. Read each scenario, and then answer each question in detail.

1. It is 12:30 in the afternoon when your engine is dispatched to a two-story apartment building. Dispatch reports that the fire is located on the second floor. The interior attack teams are fighting the fire as you arrive. The attack team reports that the fire is knocked down on the second floor but there is fire in the attic. The IC tells your Lieutenant to ladder the building and perform vertical ventilation. Your Lieutenant tells two members of your crew to ladder the building, and you and your partner to grab the chainsaw and complete a louver cut over the seat of the fire. How should you proceed?

   _____
   _____
   _____
   _____
   _____
   _____

2. It is 2:00 in the morning when your engine is dispatched to a residential structure fire. You are on the second engine to arrive on scene. The interior attack team is met with smoke and high heat at the front door. The attack team notifies the IC that ventilation is needed prior to entering the structure. The IC assigns you and your partner to coordinate positive-pressure ventilation with the attack team. How will you proceed?

   _____
   _____
   _____
   _____
   _____
   _____

Fire Fighter II in Action

The following scenario will give you an opportunity to apply your firefighting knowledge and your fire department SOGs to the new information you learned while studying this chapter. Research your department's SOGs and answer the assignment in detail. Compare your answers with your classmates' and discuss similarities and obvious differences between your answers.

You have been reading your fire magazines, and learned about several fire fighters being injured in a roof collapse while attempting vertical ventilation. Some of the injuries were severe and life- or career-threatening.

1. What are some of the indicators of a potential roof collapse?

2. What should you look for during initial size-up that may indicate potential roof collapse?

3. How do your department's SOGs on roof assignments protect your crew and you during roof operations?

Skill Drills

Skill Drill 15-1: Breaking Glass with a Hand Tool
Test your knowledge of this skill drill by filling in the correct words in the photo captions.

1. Position yourself to the _____ of the window.

2. With your back facing the wall, swing _____ forcefully with the tip of the tool striking the top one-third of the glass.

3. Clear remaining glass from the opening with the _____ _____.

Skill Drill 15-5: Delivering Positive-Pressure Ventilation

Test your knowledge of this skill drill by placing the photos below in the correct order. Number the first step with a "1," the second step with a "2," and so on.

_____ Provide an exhaust opening at or near the fire.

_____ Place the fan in front of the opening to be used for the fire attack.

_____ Start the fan and allow the smoke to clear.

Skill Drill 15-7: Sounding a Roof
Test your knowledge of this skill drill by filling in the correct words in the photo captions.

1. Use a hand tool to check the roof before _____ onto it.

2. Use the tool to sound ahead and on both _____ as you walk. Locate the support members by sound and _____. Check conditions around your work area periodically.

3. Sound the roof along your _____ path.

Skill Drill 15-9: Making a Rectangular or Square Cut

Test your knowledge of this skill drill by placing the photos below in the correct order. Number the first step with a "1," the second step with a "2," and so on.

© Jones & Bartlett Learning. Photographed by Glen E. Ellman.

_____ Pull out or push in the triangle cut.

© Jones & Bartlett Learning. Photographed by Glen E. Ellman.

_____ Make a triangle cut at the first corner.

© Jones & Bartlett Learning. Photographed by Glen E. Ellman.

_____ Locate the roof supports by sounding. Make the first cut parallel to the roof support.

© Jones & Bartlett Learning. Photographed by Glen E. Ellman.

_____ Punch out the ceiling below. Be wary of a sudden updraft of hot gases or flames.

© Jones & Bartlett Learning. Photographed by Glen E. Ellman.

_____ Make two cuts perpendicular to the roof supports and then make the final cut parallel to another roof support.

Skill Drill 15-10: Making a Louver Cut

Test your knowledge of this skill drill by filling in the correct words in the photo captions.

1. Locate the roof supports by _____.

2. Make two parallel cuts _____ to the roof supports.

3. Cut parallel to the supports and between pairs of supports in a(n) _____ pattern.

4. Tilt the _____ to a vertical position.

Skill Drill 15-11: Making a Triangular Cut

Test your knowledge of this skill drill by filling in the correct words in the photo captions.

1. Locate the roof _____.

2. Make the first cut from just _____ a support member in a(n) _____ direction toward the next support member.

3. Begin the second cut at the same location as the first, and make it in the _____ diagonal direction, forming a V shape.

4. Make the final cut _____ the support member so as to connect the first two cuts. Cutting from this location allows fire fighters the full support of the member directly below them while performing _____.

Water Supply

Workbook Activities

The following activities have been designed to help you. Your instructor may require you to complete some or all of these activities as a regular part of your fire fighter training program. You are encouraged to complete any activity that your instructor does not assign as a way to enhance your learning in the classroom.

Chapter Review

The following exercises provide an opportunity to refresh your knowledge of this chapter.

Matching

Match each of the terms in the left column to the appropriate definition in the right column.

_____ 1. Primary feeder
_____ 2. Static pressure
_____ 3. Reservoir
_____ 4. Distributor
_____ 5. Flow pressure
_____ 6. Water main
_____ 7. Shut-off valve
_____ 8. Dump valve
_____ 9. Residual pressure
_____ 10. Pitot gauge

A. Any valve that can be used to shut down water flow
B. Any underground water pipe
C. The quantity of water flowing through an opening during a hydrant test
D. A small-diameter underground water pipe, carries water to the user
E. The largest-diameter pipe in a water distribution system
F. Enables tankers to offload as much as 3000 gallons of water per minute
G. The pressure remaining in the system while water is flowing
H. A gauge used to determine the flow of water from a hydrant
I. A water storage facility
J. The pressure in a water pipe when there is no water flowing

Multiple Choice

Read each item carefully, and then select the best response.

_____ 1. What is the recommended minimum water pressure from a fire hydrant?
   A. 10 psi (69 kPa)
   B. 20 psi (138 kPa)
   C. 40 psi (276 kPa)
   D. 60 psi (414 kPa)

_____ 2. Fire department hoses can be connected to a hydrant by the
   A. valves.
   B. outlets.
   C. ports.
   D. taps.

CHAPTER 16

_____ 3. The flow or quantity of water moving through a pipe, hose, or nozzle is measured by
   A. volume.
   B. area.
   C. weight.
   D. mass.

_____ 4. Which type of hydrant includes a pipe with a strainer on one end and a connection for a hard suction hose on the other end that can be used to access static water sources?
   A. Dry hydrant
   B. Wet hydrant
   C. Static hydrant
   D. Straining hydrant

_____ 5. Which type of system may not require pumps because the water source, the treatment plant, and storage facilities are located on ground higher than the end users?
   A. Gravity-feed system
   B. Wet hydrant system
   C. Tanker shuttle system
   D. Static system

_____ 6. Hydrants should be positioned so that the connections, and especially the large steamer connection,
   A. are parallel to the water system.
   B. will not be damaged by passing traffic.
   C. are near waste storage areas.
   D. face the street.

_____ 7. What is the best indication of how much more water is available in the system while water is flowing?
   A. The elevation pressure
   B. The static pressure
   C. The residual pressure
   D. The potential pressure

_____ 8. If a large volume of water is needed for an extended period, tankers can be used to deliver water from a fill site to the scene, thereby creating
   A. a mobile water system.
   B. portable tanks.
   C. a mobile water supply apparatus.
   D. a tanker shuttle.

_____ 9. The water source, treatment plant, and distribution system are parts of a
   A. municipal water system.
   B. private water system.
   C. static water source.
   D. reservoir.

_____ 10. The pipes that deliver large quantities of water to a section of a town or city are the
   A. primary feeders.
   B. secondary feeders.
   C. direct mains.
   D. distributors.

_____ 11. To ensure that water flows to a fire hydrant from two or more directions, well-designed systems follow a
   A. mixed pattern.
   B. multiple-port pattern.
   C. grid pattern.
   D. center pattern.

_____ 12. The quantity of water flowing through an opening during a hydrant test is the
   A. residual pressure.
   B. flow pressure.
   C. normal operating pressure.
   D. elevation pressure.

_____ 13. A large opening on a fire hydrant that is used to allow as much water as possible to flow directly into the pump is a(n)
   A. outlet.
   B. steamer port.
   C. valve.
   D. drain.

_____ 14. The distribution system of underground pipes is known as
   A. reservoirs.
   B. piping.
   C. water mains.
   D. water traffic.

_____ 15. What are the first factors to check when inspecting hydrants?
   A. Stability and structural integrity
   B. Visibility and structural integrity
   C. Visibility and accessibility
   D. Stability and component location

_____ 16. The smallest pipes in a water distribution system that carry the water to the users and hydrants are the
   A. primary feeders.
   B. secondary feeders.
   C. water mains.
   D. distributors.

_____ 17. The size of water mains depends on the amount of water needed for both normal consumption and
   A. heavy consumption.
   B. extended delays.
   C. fire protection.
   D. business operations.

_____ 18. Which unit is used to measure water pressure?
   A. Gallons
   B. Gallons per square inch
   C. Pounds per square inch
   D. Pounds

_____ 19. Most dry-barrel hydrants have _____ large valve(s) controlling the flow of water.
   A. one
   B. two
   C. three
   D. four

_____ 20. Water that is not moving has
   A. elevation pressure.
   B. static pressure.
   C. residual pressure.
   D. potential pressure.

Vocabulary

Define the following terms using the space provided.

1. Municipal water system:

2. Tanker shuttle:

3. Dry-barrel hydrant:

4. Normal operating pressure:

5. Gravity-feed system:

Fill-in

Read each item carefully, and then complete the statement by filling in the missing word(s).

1. Most hydrants have an upright steel casing or _____ that is attached to the underground water distribution system.

2. The flow or quantity of water moving through a pipe is measured by its _____, usually in gallons per minute.

3. Elevation pressure can be created by _____.

4. Rural areas may depend on _____ _____ _____ such as lakes and streams.

5. Water can be transported through long hose lines, pumper relays, or _____ water supply tankers.

6. Water that is not moving has _____ energy.

7. The importance of a dependable and adequate _____ _____ for fire-suppression operations is self-evident.

8. In many communities, hydrants are painted in bright reflective colors for increased _____.

9. Municipal water systems can draw water from human-made storage facilities called _____.

10. Hydrants are equipped with one or more _____ to control the flow of water through the hydrant.

11. Dry-barrel hydrants need to be either _____ opened or _____ closed.

12. _____ valves allow different water main sections to be turned off or isolated.

True/False

If you believe the statement to be more true than false, write the letter "T" in the space provided. If you believe the statement to be more false than true, write the letter "F."

_____ 1. Volume and water pressure are synonymous.

_____ 2. Generally, water pressure ranges from 40 psi to 60 psi (276 kPa to 414 kPa) at the delivery point.

_____ 3. The Pitot gauge is used to measure flow pressure through an opening during a hydrant test.

_____ 4. The basic plan for fighting most fires depends on having an adequate supply of water.

_____ 5. Residual pressure is the amount of pressure that remains in the system when water is flowing.

_____ 6. Elevated water storage towers are used to increase the efficiency of treatment facilities.

_____ 7. Fire fighters must understand how to inspect and maintain a fire hydrant.

_____ 8. The backup water supply for some municipal systems can be large enough to store enough water for several months or years of municipal use.

_____ 9. The water department can often increase the flow of water within the system or to specific areas.

_____ 10. When the hydrant valve is opened, the drain closes in a dry-barrel hydrant.

Short Answer

Complete this section with short written answers using the space provided.

1. List the duties that need to be included in a hydrant inspection.

2. Identify the two water sources fire fighters rely on.

3. Describe the differences between dry-barrel and wet-barrel hydrants.

4. Describe the procedure a fire fighter should follow to ensure there is no foreign matter in a dry-barrel hydrant.

Fundamentals of Fire Fighter Skills

5. Describe how to perform a service test on a fire hose.

_____
_____
_____
_____
_____

6. List the information that should be noted on a hose record.

_____
_____
_____
_____
_____

7. Explain how to perform a visual hose inspection and mark a defective hose.

_____
_____
_____
_____
_____

Word Fun

The following crossword puzzle is an activity provided to reinforce correct spelling and understanding of terminology associated with firefighting. Use the clues provided to complete the puzzle. Do not include spaces or punctuation when filling in the puzzle.

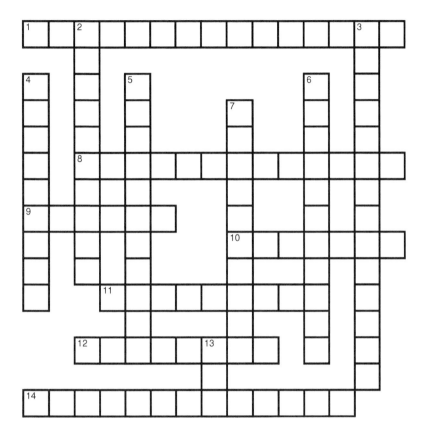

CLUES

Across

1. A hose used for drafting water from static supplies (lakes, rivers, wells, and so forth). It can also be used for supplying pumps on fire apparatus from hydrants if designed for that purpose. The hose contains a semirigid or rigid reinforcement. (NFPA 1963)
8. A type of tool used to couple or uncouple hoses by turning the rocker lugs on the connections.
9. A fungus that can grow on hose if the hose is stored wet. Mildew can damage the jacket of a hose.
10. A fitting used to connect a small hose line or pipe to a larger hose line or pipe. (NFPA 1142)
11. A device used to compress a fire hose so as to stop water flow.
12. A valved device that splits a single hose into two separate hoses, allowing each hose to be turned on and off independently.
14. Folding or collapsible tanks that are used at the fire scene to hold water for drafting.

Down

2. A method of laying a supply line where the supply line starts at the attack engine and ends at the water source.
3. The pressure that exists at a given point under normal distribution system conditions measured at the residual hydrant with no hydrants flowing. (NFPA 24)
4. A generic term for any underground water pipe.
5. The large-diameter port on a hydrant.
6. A source of water for firefighting activities. (NFPA 1144)
7. A method of laying a supply line where the line starts at the water source and ends at the attack engine.
13. A device used to split a single hose into two separate lines.

Fire Alarms

The following real case scenarios will give you an opportunity to explore the concerns associated with water supply. Read each scenario, and then answer each question in detail.

1. You have just finished lunch. Your Lieutenant has scheduled you to inspect and maintain hydrants in a residential subdivision. Your engine arrives at the first hydrant. How should you proceed?

   _____
   _____
   _____
   _____
   _____
   _____

2. It is 3:00 on an August afternoon when your tanker is dispatched to a barn fire. The first engine on the scene reports a large barn that is 75 percent involved with multiple exposures. When you arrive on the scene, you are instructed to set up a portable tank at the top of the driveway for a tanker shuttle. The next-in engine will draft from the tank and hook into the supply line. Your department has metal frame portable tanks. How should you proceed?

   _____
   _____
   _____
   _____
   _____
   _____
   _____
   _____

Fire Fighter II in Action

The following scenario will give you an opportunity to apply your firefighting knowledge and your fire department SOGs to the new information you learned while studying this chapter. Research your department's SOGs and answer the assignment in detail. Compare your answers with your classmates' and discuss similarities and obvious differences between your answers.

Your company has just finished the annual fire hose testing for your first line engine and the spare engine that is housed at your station. Several sections of fire hose failed the test and have to be repaired or scrapped.

1. How does your department determine which fire hose to repair, and which sections are destroyed?

2. What information does your department keep on each section of fire hose it possesses?

3. What information, if any, does your company keep on the fire hose on your first line and spare rigs?

Skill Drills

Skill Drill 16-2: Operating a Fire Hydrant

Test your knowledge of this skill drill by placing the photos below in the correct order. Number the first step with a "1," the second step with a "2," and so on.

© Jones & Bartlett Learning. Photographed by Glen E. Ellman.

_____ Attach the hydrant wrench to the stem nut. Check for an arrow indicating the direction to turn to open.

© Jones & Bartlett Learning. Photographed by Glen E. Ellman.

_____ Check that the remaining caps are snugly attached (dry-barrel hydrant only).

© Jones & Bartlett Learning. Photographed by Glen E. Ellman.

_____ Open the hydrant slowly to avoid a pressure surge.

© Jones & Bartlett Learning. Photographed by Glen E. Ellman.

_____ Quickly look inside the hydrant opening for foreign objects (dry-barrel hydrant only).

© Jones & Bartlett Learning. Photographed by Glen E. Ellman.

_____ Remove the cap from the outlet you will be using.

© Jones & Bartlett Learning. Photographed by Glen E. Ellman.

_____ Open the hydrant enough to verify flow and flush the hydrant (dry-barrel hydrant only).

© Jones & Bartlett Learning. Photographed by Glen E. Ellman.

_____ When instructed, turn the hydrant wrench to fully open the valve.

© Jones & Bartlett Learning. Photographed by Glen E. Ellman.

_____ Shut off the flow of water (dry-barrel hydrant only).

© Jones & Bartlett Learning. Photographed by Glen E. Ellman.

_____ Attach the hose or valve to the hydrant outlet.

Skill Drill 16-4: Testing a Fire Hydrant
Test your knowledge of this skill drill by filling in the correct words in the photo captions.

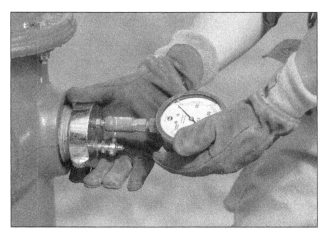

© Jones & Bartlett Learning. Photographed by Glen E. Ellman.

1. Place an _____ _____ on one of the outlets of the first hydrant.

© Jones & Bartlett Learning. Photographed by Glen E. Ellman.

2. Open the _____ _____ to fill the hydrant barrel. Record the initial pressure reading on the gauge as the _____ pressure.

© Jones & Bartlett Learning. Photographed by Glen E. Ellman.

3. Move to the second hydrant, remove one of the _____ caps, and open the hydrant.

© Jones & Bartlett Learning. Photographed by Glen E. Ellman.

4. Place the _____ _____ one-half the diameter of the orifice away from the opening and record this pressure as the _____ pressure. Record the pressure on the first hydrant as the _____ pressure. Use the recorded pressures to calculate or look up the flow rates at 20 psi (138 kPa) _____ pressure. Document your findings.

Skill Drill 16-19: Performing a Forward Hose Lay

Test your knowledge of this skill drill by filling in the correct words in the photo captions.

1. Stop the fire apparatus _____ _____ from the hydrant.

2. Grasp enough hose to reach to the hydrant and to _____ _____ the hydrant. Step off the apparatus, carrying the hydrant wrench and all necessary tools. Loop the end of the hose around the hydrant or secure the hose as specified in the local standard operating procedure (SOP). Do not stand between the hose and the hydrant. Never stand on the hose.

3. Signal the pump driver/operator to _____ to the _____ once the hose is secured.

4. Once the apparatus has moved off and a length of _____ _____ has been removed from the apparatus and is lying on the ground, remove the appropriate size _____ _____ nearest to the fire. Follow the local SOP for checking the operating condition of the hydrant.

5. Attach the _____ _____ to the _____. An adaptor may be needed if a large-diameter hose with Storz-type couplings is used.

6. Attach the _____ _____ to the hydrant.

7. The pump driver/operator uncouples the hose and attaches the end of the supply line to the _____ _____ or clamps the hose close to the pump, depending on the local SOP. When the pump driver/operator signals to charge the hose by prearranged hand signal, radio, or air horn, open the hydrant _____ and _____.

8. Follow the hose back to the _____ and remove any kinks from the supply line.

Fire Attack and Foam

Workbook Activities

The following activities have been designed to help you. Your instructor may require you to complete some or all of these activities as a regular part of your fire fighter training program. You are encouraged to complete any activity that your instructor does not assign as a way to enhance your learning in the classroom.

Chapter Review

The following exercises provide an opportunity to refresh your knowledge of this chapter.

Matching

Match each of the terms in the left column to the appropriate definition in the right column.

_____ 1. Adjustable-gallonage fog nozzle
_____ 2. Attack engine
_____ 3. Cellar nozzle
_____ 4. Hose clamp
_____ 5. Booster hose
_____ 6. Water curtain nozzle
_____ 7. Fixed-gallonage fog nozzle
_____ 8. Breakaway type nozzle
_____ 9. Wye
_____ 10. Smooth-bore nozzle
_____ 11. Fog-stream nozzle
_____ 12. Nozzles
_____ 13. Piercing nozzle
_____ 14. Protein foam
_____ 15. Aeration

A. Separates the water into droplets
B. Used to deliver a flat screen of water that then forms a protective sheet of water on the surface of an exposed building
C. A device used to compress a fire hose so as to stop water flow
D. Has a limited flow, in the range of 40 to 50 gpm
E. A nozzle that allows the operator to select a desired flow from several settings
F. Introducing air into the foam solution to produce a consistent bubble structure
G. A nozzle used to fight fires in cellars and other inaccessible places
H. Delivers a preset flow at the rated discharge pressure
I. Used to make a hole in automobile sheet metal, aircraft, or building walls so as to extinguish fires behind these surfaces
J. Can be separated between the shut-off and the tip
K. The engine from which the attack lines have been pulled
L. Capable of deeper penetration into burning materials, resulting in quicker fire knockdown and extinguishment
M. A foam that is made of animal by-products
N. A device used to split a single hose into two separate lines
O. Attachments to the discharge end of attack hoses

CHAPTER 17

Multiple Choice

Read each item carefully, and then select the best response.

_____ 1. Attack hose must be tested annually at a pressure of at least
   A. 200 psi (1379 kPa).
   B. 300 psi (2068 kPa).
   C. 400 psi (2758 kPa).
   D. 600 psi (4137 kPa).

_____ 2. Which carry is used to transport full-length attack line up a stairway?
   A. Shoulder carry
   B. Fireman's carry
   C. Hose carry
   D. Advancing carry

_____ 3. Which nozzles separate the water into droplets?
   A. Smooth-bore nozzles
   B. Fog-stream nozzles
   C. Breakaway nozzles
   D. Aeration nozzles

_____ 4. A 1 ¾-inch handline hose is generally considered to flow
   A. 60–90 gallons of water per minute.
   B. 120–180 gallons of water per minute.
   C. 250–300 gallons of water per minute.
   D. over 300 gallons of water per minute.

_____ 5. How should attack lines be loaded in the hose bed?
   A. Nozzle toward the cab
   B. Shut off toward the cab
   C. In a manner that ensures they can be quickly stretched from the attack engine to the fire
   D. Spanner toward the cab

_____ 6. Attack lines can be either double-jacket or _____ construction.
   A. rubber-covered
   B. linen
   C. vinyl
   D. quadruple jacket

_____ 7. When is a preconnected flat load ready for use?
   A. When the load is finished and the nozzle is attached
   B. When the loop is added
   C. When the female end is attached to the discharge
   D. When the adapter is added

_____ 8. Preconnected hose lines can be placed in what location(s) on a fire engine?
   A. A divided hose bed at the rear
   B. A transverse bed above the pump
   C. In the front bumper
   D. All of the above

_____ 9. The three groups of nozzles are low-volume, master-stream, and
  A. high-volume nozzles.
  B. secondary-stream nozzles.
  C. apparatus nozzles.
  D. handline nozzles.

_____ 10. A booster hose is generally considered to flow
  A. 40–50 gallons of water per minute.
  B. 60–90 gallons of water per minute.
  C. 125–150 gallons of water per minute.
  D. over 200 gallons of water per minute.

_____ 11. To perform an interior fire attack, an attack line is usually advanced in how many stages?
  A. One stage
  B. Two stages
  C. Three stages
  D. Four stages

_____ 12. Bresnan distributor nozzles are used to fight fires in
  A. warehouses.
  B. open spaces.
  C. inaccessible places.
  D. defensive attacks.

_____ 13. Which nozzle allows the operator to select a desired flow from several settings?
  A. Fixed-gallonage fog
  B. Adjustable-gallonage fog
  C. Automatic-adjusting fog
  D. Distributor

_____ 14. Which nozzle is used to make a hole in automobile sheet metal?
  A. Cellar
  B. Bresnan
  C. Piercing
  D. Smooth bore

_____ 15. Which nozzle is used on deck guns, portable monitors, and ladder pipes?
  A. Low volume
  B. Handline
  C. Master stream
  D. Straight bore

_____ 16. A 2 ½-inch handline hose is generally considered to flow
  A. 100 gallons of water per minute.
  B. 150 gallons of water per minute.
  C. 200 gallons of water per minute.
  D. 250 gallons of water per minute.

_____ 17. The end of the hose bed closest to the tailboard is called the
  A. transverse.
  B. front.
  C. rear.
  D. bumper.

_____ 18. Which foam application method uses an object to deflect the foam stream down onto the fire?
  A. Overhead method
  B. Deflection method
  C. Bankshot method
  D. Rain-down method

Chapter 17: Fire Attack and Foam

_____ 19. Which type of foam is used to fight fires involving ordinary combustible materials?
   A. Class A foam
   B. Class B foam
   C. Protein foam
   D. Fluoroprotein foam

_____ 20. Which device mixes the foam concentrate into the fire stream?
   A. Foam eductor
   B. Foam injector
   C. Foam regulator
   D. Foam proportioner

_____ 21. Which is a correct foam application technique?
   A. Sweep
   B. Bankshot
   C. Rain-down
   D. All of the above

_____ 22. How can foam be applied to a fire or spill?
   A. Portable extinguishers
   B. Handlines
   C. Master streams
   D. All of the above

_____ 23. What device adds the foam concentrate to the water stream under pressure?
   A. Foam injectors
   B. Aerator
   C. Bresnan nozzle
   D. Pick-up tube

_____ 24. Which of the following is not a Class B foam concentrate?
   A. Protein
   B. AFFF
   C. FFFP
   D. CAFS

_____ 25. Once a foam blanket has been applied
   A. you may walk through the spill.
   B. keep a fog spray on the foam surface.
   C. it must not be disturbed.
   D. the fuel under the foam cannot produce flammable vapors.

Vocabulary

Define the following terms using the space provided.

1. Handline nozzle:

   _____
   _____
   _____

2. Smooth-bore nozzle:

   _____
   _____
   _____

3. Fixed-gallonage fog nozzle:

   _____
   _____
   _____

4. Foam proportioners:

   _____
   _____
   _____

5. Batch mixing:

   _____
   _____
   _____

Fill-in

Read each item carefully, and then complete the statement by filling in the missing word(s).

1. Attack lines should be loaded in the hose bed in a manner that ensures they can be quickly stretched from the _____ _____ to the fire.

2. _____-_____ _____ _____ are often used for fighting wildland and ground fires.

3. A(n) _____ _____ is usually carried on a hose reel that holds 150 or 200 feet (46 to 61 meters) of rubber hose.

4. Large-diameter hose (LDH) has a limited role as a(n) _____ _____ tool.

5. Hose should be _____ _____ with lengths of hose running parallel to the front of the fire building so that it can be easily advanced into the building.

6. If a hose line has to be advanced up a ladder, it should be positioned correctly before the line is _____.

7. When you advance a hose line down a stairway, you have a major advantage on your side: _____.

8. When advancing an attack line up a ladder, once the first fire fighter reaches the _____ _____ _____ of the ladder, a second fire fighter shoulders the hose to assist advancing the hose line up the ladder.

9. The hoses used to discharge water from an attack engine onto the fire are called _____ _____.

10. Once hose is flaked out, signal the driver/operator to _____ _____ _____.

11. When you are given the command to advance the hose, keep _____ as your number one priority.

12. Ideally, a hose line crew consists of at least three members at the nozzle and a fourth member _____ _____ _____.

13. Fire fighters can add hose to the discharge end of the hose if the nozzle is a(n) _____ _____ nozzle.

14. Standpipe outlets are often located in stairways, and standard operating procedures (SOPs) generally require attack lines to be connected to an outlet _____ _____ _____ the fire.

15. A _____ fire attack is an aggressive offensive exterior attack that occurs in coordination with entry search and tactical ventilation.

16. Hoses containing _____ tend to kink more easily than hoses filled with water.

True/False

If you believe the statement to be more true than false, write the letter "T" in the space provided. If you believe the statement to be more false than true, write the letter "F."

_____ 1. The straight stream from a fog-stream nozzle breaks up faster and does not have the reach of a solid stream.

_____ 2. The advantage of booster hose is its large flow.

_____ 3. Laying out fire hose should not require multiple trips between the engine and the fire building.

_____ 4. When the attack line has been laid out to the entry point, the extra hose that will be advanced into the building should be flaked out in a serpentine pattern.

_____ 5. Charged hose lines are easy to advance through a house or other building.

_____ 6. When advancing a hose line up stairs, arrange to have an adequate amount of extra hose close to the pump panel.

_____ 7. The triple-layer load requires several fire fighters and must be practiced often.

_____ 8. Additional fire fighters should pick up the hose about every 50 feet (15 meters) and help to advance it up a ladder.

_____ 9. When choosing a preconnected hose line or assembling an attack line, it is better to have too much hose than not enough.

_____ 10. When placing a pickup tube from an eductor into the foam concentrate, keep items at similar elevations.

_____ 11. Some vehicles are equipped with large tanks holding 50 or 100 gallons (189 or 379 liters) of premixed foam.

_____ 12. Foam can be applied with a wide range of expansion rates, depending on the amount of air that is mixed into the stream and the size of the bubbles that are produced.

_____ 13. Medium-expansion foam is produced with special aerating nozzles that are designed to introduce less air into the stream and produce an inconsistent bubble structure.

_____ 14. When attempting to use foam to extinguish a flammable-liquid fire, it is not necessary to cover the spill completely.

_____ 15. Hose lines containing CAF are noticeably lighter than lines completely filled with water.

_____ 16. When used for interior firefighting, drier foam is more appropriate.

Short Answer

Complete this section with short written answers using the space provided.

1. Describe how foam suppresses fire.

2. Describe the characteristics of Class A foam.

3. Describe the characteristics of Class B foam.

4. List the major categories of Class B foam concentrate.

5. Describe the characteristics of the following:

   A. Protein foam

   B. Fluoroprotein foam

   C. Aqueous film-forming foam

   D. Alcohol-resistant foam

   E. Compressed air foam

6. Describe how foam proportioner equipment works with foam concentrate to produce foam.

7. Describe how foam is applied to fires.

Fire Alarms

The following real case scenarios will give you an opportunity to explore the concerns associated with fire hoses, nozzles, streams, and foam. Read each scenario, and then answer each question in detail.

1. Your new engine company carries its attack lines preconnected in a minuteman hose load. Your company officer orders you to demonstrate loading a 150-foot, 1 ¾-inch minuteman hose load. How will you proceed?

2. Your company officer orders you to demonstrate advancing a 1 ¾-inch minuteman hose load. Explain this procedure in detail.

Fundamentals of Fire Fighter Skills

Fire Fighter II in Action

The following scenario will give you an opportunity to apply your firefighting knowledge and your fire department SOGs to the new information you learned while studying this chapter. Research your department's SOGs and answer the assignment in detail. Compare your answers with your classmates' and discuss similarities and obvious differences between your answers.

Engines carry a variety of sizes of fire hose, as well as a number of different size and type nozzles. Obviously, there are reasons for the variety.

1. List the different hose sizes your department carries. Identify the applications for each size, and also identify situations when it would not be advisable to use each size.

   _____
   _____
   _____
   _____
   _____
   _____
   _____

2. List the different nozzles your department carries. Identify the applications and advantages of each nozzle.

   _____
   _____
   _____
   _____
   _____
   _____
   _____

Skill Drills

Skill Drill 17-10: Advancing an Uncharged Hose Line up a Ladder
Test your knowledge of this skill drill by filling in the correct words in the photo captions.

1. Advance the hose line to the ladder. Pick up the _____; place the hose across the chest, with the nozzle draped over the _____. Climb up the ladder with the _____ hose line.

2. Once the first fire fighter reaches the first fly section of the ladder, a second fire fighter shoulders the hose to assist advancing the hose line up the ladder. To avoid overloading of the ladder, enforce a limit of _____ fire fighter(s) per fly section. The nozzle is placed over the _____ _____ of the ladder and advanced into the _____ _____.

3. Additional hose can be fed up the ladder until sufficient hose is in position. The hose can be secured to the ladder with a(n) _____ _____ to support its weight and keep it from becoming dislodged.

Fire Fighter Survival

Workbook Activities

The following activities have been designed to help you. Your instructor may require you to complete some or all of these activities as a regular part of your fire fighter training program. You are encouraged to complete any activity that your instructor does not assign as a way to enhance your learning in the classroom.

Chapter Review

The following exercises provide an opportunity to refresh your knowledge of this chapter.

Matching

Match each of the terms in the left column to the appropriate definition in the right column.

_____ 1. Hazardous conditions   **A.** Indicates a fire fighter is in trouble and requires immediate assistance

_____ 2. Guideline   **B.** A rope used for orientation when inside a structure when there is low or no visibility

_____ 3. SOPs   **C.** Defines the manner in which a fire department conducts operations at an emergency incident

_____ 4. Safe location   **D.** An extension of the two-in/two-out rule

_____ 5. Self-rescue   **E.** Becomes easier through study and experience

_____ 6. Hazard recognition   **F.** May not be evident by simple observation

_____ 7. Rehabilitation   **G.** A temporary place of refuge in which to await rescue

_____ 8. Risk–benefit analysis   **H.** Weighs the positive results that can be achieved against the probability and severity of potential negative consequences

_____ 9. RIC   **I.** Escaping or exiting a hazardous area under one's own power

_____ 10. Mayday   **J.** Reduces the effects of fatigue during an emergency operation

Multiple Choice

Read each item carefully, and then select the best response.

_____ 1. When a fire fighter needs immediate assistance, the incident commander should immediately deploy
     A. the SOP.
     B. the PASS.
     C. the RIC.
     D. the radio.

_____ 2. Fire fighters will accept a higher level of risk in exchange for
     A. the possibility of saving lives.
     B. the possibility of saving property.
     C. property that is lost.
     D. persons who are already lost.

CHAPTER 18

_____ 3. When initiating a mayday,
   A. give a weather report.
   B. give a LUNAR report.
   C. give a CISD report.
   D. give a PASS report.

_____ 4. Upon reaching a downed fire fighter, what is the most critical decision for the rescuers?
   A. How much time and effort will be needed to remove the fire fighter
   B. The treatment of the fire fighter's injuries
   C. The location of the fire fighter and rescuers
   D. How to exit the structure

_____ 5. Which of the following must be learned and practiced before they can be implemented?
   A. CISMs
   B. SOPs
   C. GOPs
   D. DUIs

_____ 6. Critical incident stress that is cumulative, building up over time, is called
   A. normal.
   B. remorse.
   C. burnout.
   D. self-reconciliation.

_____ 7. Observable factors that might indicate a hazard include
   A. building construction.
   B. weather conditions.
   C. occupancy.
   D. all of the above.

_____ 8. What does the NFPA 704 diamond indicate?
   A. PARs are present.
   B. RICs are present.
   C. Ventilation is necessary.
   D. Hazardous materials are present.

_____ 9. The recognized stages of emotional reaction experienced by fire fighters and other rescue personnel after a stressful incident can include
   A. anxiety.
   B. denial/disbelief.
   C. frustration/anger.
   D. all of the above.

_____ 10. What is the most important outcome in any fire department operation?
   A. Fire fighter survival
   B. No rekindle
   C. Effective ICS
   D. Reduced water damage

_____ 11. Comparing potential positive results to potential negative consequences is called
   A. causative factors.
   B. management factors.
   C. risk–benefit analysis.
   D. standard operating procedures.

_____ 12. To stay oriented when inside a burning structure, the fire fighter should use a hose line or a
   A. team member.
   B. radio.
   C. guideline.
   D. structure wall.

_____ 13. A roll call taken by each supervisor at an emergency incident is known as a(n)
   A. team roll call.
   B. incident report.
   C. personnel accountability report.
   D. incident roll.

_____ 14. While awaiting rescue, a fire fighter may find a temporary location that provides refuge. What is this location called?
   A. Safety point
   B. Safe location
   C. Rescue point
   D. Landmark

_____ 15. What is the first step of self-rescue?
   A. Manually set off your PASS alarm.
   B. Initiate a mayday.
   C. Exit the structure.
   D. Orient yourself within the structure.

_____ 16. During an incident, if fire fighters observe an increase in risk of their operations, they must report it to the
   A. company officer.
   B. sector officer.
   C. safety officer.
   D. incident commander.

_____ 17. The manner in which a fire department conducts operations at an emergency incident is defined by
   A. general operating guidelines.
   B. the incident commander.
   C. department policies.
   D. standard operating procedures.

_____ 18. The standard radio terminology used to report an imminent hazardous condition or situation is
   A. "Emergency traffic."
   B. "Mayday."
   C. "Halt operations."
   D. "Retreat."

_____ 19. A crew that is assigned to stand by fully dressed, equipped for action, and ready to deploy at an incident scene is called a(n)
   A. technical rescue crew.
   B. EMS team.
   C. special recovery team.
   D. rapid intervention crew.

_____ 20. A systematic way to keep track of the location and function of all personnel operating at the scene of an incident is
   A. a team inventory.
   B. a personnel accountability system.
   C. the chain of command.
   D. the two-in/two-out rule.

Vocabulary

Define the following terms using the space provided.

1. Critical incident stress management (CISM):

    _____
    _____
    _____

2. Safe location:

    _____
    _____
    _____

3. Air management:

    _____
    _____
    _____

4. Self-rescue:

    _____
    _____
    _____

5. Rapid intervention company/crew (RIC):

    _____
    _____
    _____

Fill-in

Read each item carefully, and then complete the statement by filling in the missing word(s).

1. Hazardous conditions may or may not be evident by _____ _____.
2. The word _____ is used to indicate that a fire fighter is in trouble and requires immediate assistance.
3. During an incident, company officers and safety officers are involved in risk analysis on a(n) _____ basis.
4. Air _____ is important to all fire fighters and relates to the basic fact that air equals time.
5. The assessment of the risks and benefits and the decision to commit crews to the interior of a burning structure is the responsibility of the _____ _____.
6. The only way to become proficient at a skill is through _____.
7. The purpose of _____ is to reduce the effects of fatigue during an emergency operation.
8. Team _____ means that a company arrives at a fire together, works together, and leaves together.
9. Safety and survival inside a fire building can be directly related to remaining _____ within the building.

True/False

If you believe the statement to be more true than false, write the letter "T" in the space provided. If you believe the statement to be more false than true, write the letter "F."

_____ 1. Teamwork and communication are critical parts of all emergency operations.
_____ 2. The stages of emotional reaction after a critical incident can occur within minutes or months of the incident.
_____ 3. A room with a door and a window could be used as a safe location.
_____ 4. Fire fighters must be capable of working in environments that include a wide range of hazards.
_____ 5. Critical incident stress is created by an event that is interpreted by the fire fighter as traumatic.
_____ 6. Rapid intervention crews/companies should be in place at any incident where fire fighters are in operation.
_____ 7. Sometimes fire fighters react to critical incidents in ways that are not positive.
_____ 8. The best method to remain oriented within an involved structure is to stay in contact with a team member.
_____ 9. The members of a company should always be oriented to one another's location, activities, and condition.
_____ 10. It is permissible to risk the life of a fire fighter only in a situation where there is a reasonable and realistic possibility of saving a life.

Short Answer
Complete this section with short written answers using the space provided.

1. List the recognized stages of emotional reaction experienced by fire fighters and other rescue personnel to critical incidents.

2. Identify a simply stated risk–benefit philosophy for a fire department.

3. Describe the procedure for initiating a mayday.

Fundamentals of Fire Fighter Skills

Clues

Across

5 A component integrated within the protective coat element to aid in the rescue of an incapacitated fire fighter. (NFPA 1851)

7 The use of a limited air supply in such a way as to ensure that it will last long enough to enter a hazardous area, accomplish needed tasks, and return safely.

Down

1 Escaping or exiting a hazardous area under one's own power. (NFPA 1006)

2 A rope used for orientation when fire fighters are inside a structure where there is low or no visibility.

3 A location remote or separated from the effects of a fire so that such effects no longer pose a threat. (NFPA 101)

4 Periodic report verifying the status of responders assigned to an incident or planned event. (NFPA 1026)

6 A program designed to reduce both acute and chronic effects of stress related to job functions. (NFPA 450)

Word Fun

The following crossword puzzle is an activity provided to reinforce correct spelling and understanding of terminology associated with firefighting. Use the clues provided to complete the puzzle. Do not include spaces or punctuation when filling in the puzzle.

Fire Alarms

The following real case scenarios will give you an opportunity to explore the concerns associated with fire fighter survival. Read each scenario, then answer each question in detail.

1. While performing a search in a commercial structure, you crawl into a portion of fallen suspended ceiling and become entangled in electric wire and communication cables. What steps will you take to remove yourself from the wires and cables?

2. You have become confused and disoriented during a primary search of a second-floor apartment. You decide to attempt to locate a window for calling for assistance or exit. How will you proceed?

Fire Fighter II in Action

The following scenario will give you an opportunity to apply your firefighting knowledge and your fire department's SOGs to the new information you learned while studying this chapter. Research your department's SOGs and answer the assignment in detail. Compare your answers with your classmates' and discuss similarities and obvious differences between your answers.

Year after year, fire fighters die in structure fires because they fail to realize they are in danger, or fail to react in a timely fashion. The International Association of Fire Chiefs (IAFC) and National Volunteer Fire Council (NVFC) developed the Rules of Engagement for Fire Fighter Survival. These rules are summarized in the Fire Fighter Survival chapter of your manual.

1. Following is a Web site you can access for additional information. Read these rules and incorporate them into your training and structural firefighting SOPs. http://www.iafcsafety.org/Rules_of_Engagement_v8_7.10.pdf

Skill Drills

Skill Drill 18-1: Initiating a Mayday Call
Test your knowledge of this skill drill by filling in the correct words in the photo captions.

1. Use your radio to call, "Mayday, mayday, mayday." Give a LUNAR report: your location, _____ _____, name, assignment, and _____ needed.

2. Activate your _____ device. Attempt _____-_____. If you are able to move, identify a(n) _____ _____ where you can await rescue.

3. Lie on your side in a(n) _____ position with your _____ device pointing out so it can be heard.

4. Point your flashlight toward the ceiling. Slow your breathing as much as possible to conserve your _____ _____.

Skill Drill 18-7: Rescuing a Downed Fire Fighter Using a Drag Rescue Device

Test your knowledge of this skill drill by placing the photos below in the correct order. Number the first step with a "1," the second step with a "2," and so on.

_____ Locate the downed fire fighter. Activate the mayday procedure, if that step had not already been taken. Shut off the PASS device to aid in communication. Assess the situation and the condition of the downed fire fighter.

_____ Remove the downed fire fighter from the hazard area to a safe area.

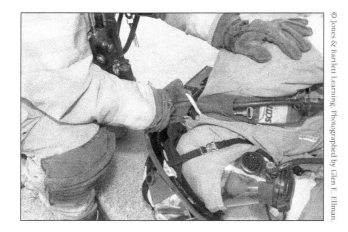

_____ Access the fire fighter's drag rescue device.

_____ Use the rapid intervention crew/company universal air connection (RIC–UAC) to fill the downed fire fighter's air supply cylinder, if needed.

Salvage and Overhaul

Workbook Activities

The following activities have been designed to help you. Your instructor may require you to complete some or all of these activities as a regular part of your fire fighter training program. You are encouraged to complete any activity that your instructor does not assign as a way to enhance your learning in the classroom.

Chapter Review

The following exercises provide an opportunity to refresh your knowledge of this chapter.

Matching

Match each of the terms in the left column to the appropriate definition in the right column.

_____ 1. Rekindle
_____ 2. Floodlight
_____ 3. Carryall
_____ 4. Water chute
_____ 5. Salvage
_____ 6. Spotlight
_____ 7. Salvage covers
_____ 8. Water catch-all
_____ 9. Generator
_____ 10. Inverter

A. Protect property and belongings from damage, particularly from the effects of smoke and water
B. Catches dripping water and directs it toward a drain or to the outside
C. A light designed to project a narrow, concentrated beam of light
D. Causes more property damage to occur, and the fire department could be held responsible for the additional losses
E. Projects a more diffuse light over a wide area
F. An electromechanical device for the production of electricity
G. Large square or rectangular sheets of heavy canvas or plastic material that are used to protect furniture and other items from water runoff, falling debris, soot, and particulate matter in smoke residue
H. A device that converts the direct current from an apparatus electrical system into alternating current
I. A temporary pond that holds dripping water in one location
J. Heavy canvas with handles used to carry debris

Multiple Choice

Read each item carefully, and then select the best response.

_____ 1. Before a fire fighter can work without a self-contained breathing apparatus (SCBA), the atmosphere must be tested and determined safe by the
   A. safety officer.
   B. incident commander.
   C. RIC leader.
   D. department captain.

CHAPTER 19

_____ 2. The building's construction, its contents, and the size of the fire are factors in determining
   A. salvage operations.
   B. the area that needs to be overhauled.
   C. who the incident commander is.
   D. the placement of the rapid intervention company/crew.

_____ 3. Sprinklers should be shut down as soon as the IC declares the fire
   A. is under control.
   B. is out.
   C. has spread to other rooms of the structure.
   D. is being ventilated.

_____ 4. Fire damage to the building's structural components may potentially lead to
   A. atmospheric contamination.
   B. reignition.
   C. rekindling.
   D. structural collapse.

_____ 5. Which type of opening drains water from an above-ground floor through an exterior wall hole?
   A. Water chute
   B. Water catch-all
   C. Drain
   D. Fire-fighter-made opening

_____ 6. What is the best way to prevent water damage at a fire scene?
   A. Increase the number of ventilation sites
   B. Use higher flow rates
   C. Use the building sprinkler system
   D. Limit the amount of water used

_____ 7. Which type of lights are most often used during the first few critical minutes of an incident?
   A. 1500-watt portable lights
   B. Battery-powered lights
   C. 300-watt quartz lights
   D. 1000-watt halogen lights

_____ 8. The main control for a sprinkler system is usually a(n)
   A. sprinkler box.
   B. scupper.
   C. OS&Y valve or PIV.
   D. water valve.

_____ 9. A long section of protective material used to cover a section of carpet is called a
   A. floor cover.
   B. floor runner.
   C. carpet tarp.
   D. drop tarp.

_____ 10. Mobile power outlets, which are placed in convenient locations for cords to be attached, are also known as
   A. junction boxes.
   B. generators.
   C. inverters.
   D. extension outlets.

_____ 11. To stop the flow from a sprinkler, insert a
   A. hand tool.
   B. sprinkler wedge.
   C. sprinkler key.
   D. cloth.

_____ 12. During salvage operations, smaller pictures and valuable objects should be placed in
   A. the pockets of the fire fighter's PPE.
   B. smaller tarps.
   C. drawers.
   D. the corner.

_____ 13. Fire can extend directly from the basement to the attic, without obvious signs of fire, in a
   A. balloon-frame building.
   B. platform-frame building.
   C. Type I building.
   D. remodeled building.

_____ 14. Efforts to protect property and belongings from damage are called
   A. overhaul.
   B. salvage.
   C. rescue.
   D. recovery.

_____ 15. The most common method of protecting building contents is to cover them with
   A. heat-reflective blankets.
   B. salvage covers.
   C. overhaul tarps.
   D. salvage tarps.

_____ 16. Fire fighters' salvage efforts at residential fires often focus on protecting
   A. expensive items or property.
   B. isolated property.
   C. personal property.
   D. market items.

_____ 17. The most efficient way to protect a room's contents is to move all the furniture to
   A. the center of the room.
   B. the wall farthest from the flames.
   C. the walls nearest the window(s) used for ventilation and fire suppression.
   D. the front of the room.

_____ 18. How much does a gallon (4 liters) of water weigh?
   A. 2.24 pounds (1 kilogram)
   B. 4.5 pounds (2 kilograms)
   C. 6 pounds (2.7 kilograms)
   D. 8.3 pounds (3.8 kilograms)

_____ **19.** How long should gasoline-powered generators be run to reduce deposit build-ups?
  A. 2 hours
  B. 1 hour
  C. 15–30 minutes
  D. 5–10 minutes

_____ **20.** To convert 12-volt DC current to 110-volt AC current, you would use a(n)
  A. power inverter.
  B. junction box.
  C. power outlet.
  D. extension.

Vocabulary

Define the following terms using the space provided.

1. Salvage cover:

2. Sprinkler wedge:

3. Floor runner:

4. Overhaul:

5. Sprinkler stop:

6. Balloon-frame construction:

Fill-in
Read each item carefully, and then complete the statement by filling in the missing word(s).

1. A(n) _____ _____ should always be present during overhaul operations to note any hazards and ensure that operations are conducted safely.

2. _____ _____ teams remain at the fire scene and watch for signs of rekindling.

3. A(n) _____ vacuum is a special piece of equipment used to suck up water during salvage operations.

4. Buckets, tubs, wheelbarrows, and _____ can be used to remove debris from a building.

5. Salvage and overhaul have a(n) _____ priority than search and rescue.

6. A(n) _____ _____ _____ can distinguish between objects or areas with different temperatures.

7. _____ ensures that a fire is completely extinguished.

8. During salvage and overhaul efforts, fire fighters must attempt to preserve _____ related to the cause of the fire, particularly when arson is expected.

9. Look, listen, and _____ to detect signs of potential burning.

10. Salvage efforts are usually aimed at preventing or limiting _____ _____ that result from smoke and water damage.

11. Sprinkler heads that have been activated must be _____ before they can be restored to normal operations.

12. Generators should have _____-_____ _____ (_____) to prevent a fire fighter from receiving a potentially fatal electric shock.

13. When cleaning and maintaining a generator, follow the _____ _____.

14. Clean the external parts of the generator with a(n) _____ _____ or _____ as recommended by the manufacturer.

15. Test and run generators on a(n) _____ or _____ basis to confirm that they will start, run smoothly, and produce power.

True/False

If you believe the statement to be more true than false, write the letter "T" in the space provided. If you believe the statement to be more false than true, write the letter "F."

_____ 1. The entire area around a fire building should be illuminated.

_____ 2. When entering an area for salvage operations, fire fighters should roll a floor runner ahead of themselves.

_____ 3. Often the damage caused to property by smoke and water can be more extensive and costly to repair or replace than the property that is burned.

_____ 4. The cause of the fire can also indicate the extent of overhaul necessary.

_____ 5. Pike poles are used to pull down sections of ceiling.

_____ 6. Salvage efforts can be done during fire suppression.

_____ 7. Sprinkler control valves should always be locked in the closed position.

_____ 8. The IC may order hydraulic or standard overhaul procedures.

_____ 9. Water used in fire suppression can create potential hazards for fire fighters.

_____ 10. Salvage crews begin on the floor of the fire to prevent water damage to room contents.

_____ 11. Gasoline-powered generators should be run for 15 to 30 minutes to reduce any deposit build-up that could foul the spark plugs and make the generator hard to start.

_____ 12. Permanently mounted generators can be powered by the apparatus engine through a power take-off and a hydraulic pump.

Short Answer

Complete this section with short written answers using the space provided.

1. List four potential hazards present during overhaul operations.

2. Identify five tools used in salvage operations.

3. List five tools used in overhaul operations.

4. List five indicators of possible structural collapse.

5. List the steps for conducting a weekly/monthly generator test:

Word Fun

The following crossword puzzle is an activity provided to reinforce correct spelling and understanding of terminology associated with firefighting. Use the clues provided to complete the puzzle. Do not include spaces or punctuation when filling in the puzzle.

CLUES

Across

3. A light designed to project a narrow, concentrated beam of light.
10. Equipment that is used to change the voltage level and/or waveform of electrical energy.
11. A salvage cover that has been folded to form a container to hold water until it can be removed.
12. A firefighting procedure for protecting property from further loss following an aircraft accident or fire. (NFPA 402)
13. A device similar to a wet/dry shop vacuum cleaner that can pick up liquids. It is used to remove water from buildings.

Down

1. A piece of wedge-shaped wood placed between the deflector and the orifice of a sprinkler head to stop the flow of water.
2. A light that can illuminate a broad area.
4. A sprinkler control valve with an indicator that reads either open or shut depending on its position.
5. A piece of canvas or plastic material, usually 3 to 4 feet (91 to 122 centimeters) wide and available in various lengths, that is used to protect flooring from dropped debris and dirt from shoes and boots.
6. An electromechanical device for the production of electricity. (NFPA 1901)
7. A piece of heavy canvas with handles, which can be used to tote debris, ash, embers, and burning materials out of a structure.
8. A salvage cover that has been folded to direct water flow out of a building or away from sensitive items or areas.
9. The process of final extinguishment after the main body of a fire has been knocked down. All traces of fire must be extinguished at this time. (NFPA 402)

Fundamentals of Fire Fighter Skills

Fire Alarms

The following real case scenarios will give you an opportunity to explore the concerns associated with salvage and overhaul. Read each scenario, and then answer each question in detail.

You are assigned to the second ladder company arriving at a third-floor apartment fire in a multistory residential structure. The IC has ordered your company to the floor below the fire for salvage operations. Water is running from the ceiling into a second-floor apartment, causing much water damage. Your company officer orders you to protect the contents of the family room and then build a water chute to divert water to an outside window.

1. Which actions will you take to protect the family room contents?

2. How will you build a water chute to divert the water from the ceiling?

3. After you return from the fire described in questions 1 and 2, your company officer instructs you to clean and check the salvage covers for wear. What is the proper method for cleaning and inspecting salvage covers?

4. After the salvage covers have dried, you are to fold them for future use. How will you fold a salvage cover so it may be deployed by a single fire fighter?

Fire Fighter II in Action

The following scenario will give you an opportunity to apply your firefighting knowledge and your fire department SOGs to the new information you learned while studying this chapter. Research your department's SOGs and answer the assignment in detail. Compare your answers with your classmates' and discuss similarities and obvious differences between your answers.

All gasoline-powered tools and equipment should be started and run weekly or monthly. This will keep them in top operating condition, be more reliable in an emergency, and keep you, the fire fighter, familiar with startup and maintenance of the equipment.

1. Read through the manufacturer's instruction manual for each of your gas-powered tools. Be sure you cover startup, maintenance, and safety information. Cross-reference these instructions with your gas-powered tool SOGs and the company recordkeeping procedures. If the SOGs need to be updated, bring them to your company officer's attention.

Skill Drills

Skill Drill 19-2: Conducting a Weekly/Monthly Generator Test
Test your knowledge of this skill drill by filling in the correct words in the photo captions.

1. Remove the generator from the _____ _____ or open all doors as needed for _____. Install the _____ _____, if needed.

2. Check the oil and fuel levels, and start the generator. Connect the power cord or _____ _____ to the generator; connect a load such as a fan or lights, and make sure the generator attains the proper _____. Check the _____ and _____ gauges to confirm efficient operation.

3. Run the generator under load for _____ to _____ minutes. Turn off the load and listen as the generator slows down to idle speed. Allow the generator to idle for approximately _____ minutes before turning it off. Disconnect all power cords and junction boxes; clean all power cords, plugs, adaptors, GFIs, and tools, and replace them in proper storage areas. Allow the generator to cool for _____ minutes. Refill the generator with fuel and oil as needed, and return the generator to its compartment. Fill out the appropriate paperwork.

Skill Drill 19-10: Performing a Salvage Cover Fold for Two-Fire Fighter Deployment

Test your knowledge of this skill drill by filling in the correct words in the photo captions.

© Jones & Bartlett Learning. Photographed by Glen E. Ellman.

1. Spread the salvage cover flat on the ground with a partner facing you. Together, fold the cover in _____.

© Jones & Bartlett Learning. Photographed by Glen E. Ellman.

2. Together, grasp the _____ edge and fold the cover in _____ again. Flatten the salvage cover to remove any _____ _____.

© Jones & Bartlett Learning. Photographed by Glen E. Ellman.

3. Move to the newly created narrow ends of the salvage cover and fold the salvage cover in half _____.

© Jones & Bartlett Learning. Photographed by Glen E. Ellman.

4. Fold the salvage cover in half _____ again. Make certain that the open end is _____ _____.

© Jones & Bartlett Learning. Photographed by Glen E. Ellman.

5. Fold the cover in _____ a third time.

Fire Fighter Rehabilitation

Workbook Activities

The following activities have been designed to help you. Your instructor may require you to complete some or all of these activities as a regular part of your fire fighter training program. You are encouraged to complete any activity that your instructor does not assign as a way to enhance your learning in the classroom.

Chapter Review

The following exercises provide an opportunity to refresh your knowledge of this chapter.

Matching

Match each of the terms in the left column to the appropriate definition in the right column.

_____ 1. Electrolytes
_____ 2. Rehabilitate
_____ 3. Frostbite
_____ 4. Fully encapsulated suits
_____ 5. Emergency incident rehabilitation
_____ 6. Glucose
_____ 7. Dehydration
_____ 8. Personal protective equipment (PPE)
_____ 9. Hypothermia
_____ 10. Rehydrating

A. A protective suit that completely covers the fire fighter, including the breathing apparatus
B. Damage to tissue resulting from exposure to cold
C. Part of the overall emergency effort
D. A condition in which the body temperature falls below 95°F (35°C)
E. Certain salts and other chemicals that are dissolved in body fluids and cells
F. A state in which fluid losses are greater than fluid intake into the body
G. To restore to a condition of health or to a state of useful and constructive activity
H. More complicated than just drinking a lot of fluids
I. Protective clothing and breathing apparatus used by fire fighters to reduce and prevent injuries
J. Needed to burn fat efficiently and release energy

Multiple Choice

Read each item carefully, and then select the best response.

_____ 1. Which of the following is *not* a function of an emergency incident rehabilitation center?
   A. Relief from climatic conditions
   B. Medical monitoring
   C. Rehydration
   D. Staging

CHAPTER 20

_____ 2. Which of the following is the best rehabilitation food source during a short duration incident?
   A. Water and a high-protein sports bar
   B. Coffee or tea and a high-protein sports bar
   C. Fruit juice and pasta
   D. A complete meal that includes complex carbohydrates

_____ 3. Conditioning plays a significant role in
   A. endurance.
   B. glucose intake.
   C. intensity.
   D. reassignment.

_____ 4. Which of the following is the only fuel that the body can readily use during high-intensity physical activity?
   A. Carbohydrates
   B. Fats
   C. Proteins
   D. Electrolytes

_____ 5. The best way to prepare for physical work is to do activities that match
   A. the type of work.
   B. the duration of work.
   C. the intensity of work.
   D. all of the above.

_____ 6. A fire fighter must stay at the rehabilitation center until his or her body temperature returns to
   A. a range consistent with the environment.
   B. a temperature above pre-entry levels.
   C. a normal range.
   D. below pre-entry levels.

_____ 7. Which of the following may require rehabilitation?
   A. Training exercises
   B. Athletic events
   C. Standby assignments
   D. All of the above

_____ 8. What is a benefit of proper nutrition?
   A. Stress reduction
   B. Health improvements
   C. Increased energy
   D. All of the above

_____ 9. Rehabilitation enables fire fighters to
   A. perform more safely and effectively.
   B. review personal protective equipment items.
   C. discuss emergency response strategies.
   D. observe and assess response tactics.

Fundamentals of Fire Fighter Skills

_____ **10.** During a high-rise fire, a department may assign three companies to do the work that is normally done by one company because
   A. there is a larger structure to protect.
   B. the higher the structure, the more distance to the ground.
   C. there is more equipment to carry.
   D. it enables the companies to rotate duties.

Vocabulary

Define the following terms using the space provided.

1. Rehabilitate:

2. Dehydration:

3. Hypothermia:

4. Frostbite:

5. Emergency incident rehabilitation:

Fill-in

Read each item carefully, and then complete the statement by filling in the missing word(s).

1. Blood sugar, or _____, is the fuel the body uses for energy.

2. When a fire fighter has abnormal vital signs, is suffering pain, or is injured, he or she needs to have further _____ _____ and _____.

3. _____ can actually cause a swing in energy levels because it stimulates the production of insulin.

4. _____ is a critical factor in maintaining health and well-being.

5. _____ happens only when a fire fighter is rested, rehydrated, refueled, and fit to return to active duty.

6. _____ is caused by drinking too much too quickly.

7. The amount of rest needed to recover from physical exertion is directly related to the _____ of the work performed.

8. The stomach can absorb about _____ quart(s) of fluid per hour, but the body can lose up to _____ quart(s) of fluid per hour.

9. _____ occurs when body tissues are damaged due to prolonged exposure to the cold.

10. _____ are a major source of fuel for the body and can be found in grains, vegetables, and fruits.

True/False

If you believe the statement to be more true than false, write the letter "T" in the space provided. If you believe the statement to be more false than true, write the letter "F."

_____ 1. Fats are used for energy and for breaking down some vitamins.

_____ 2. Regular rehabilitation enables fire fighters to accomplish more work during a major incident.

_____ 3. Rehabilitation helps improve the quality of decision making.

_____ 4. Taking short breaks, replacing fluids, ingesting healthy food, and cooling or rewarming are all measures that reduce the risk of injury and illness.

_____ 5. Rehabilitation enables fire fighters to perform more safely and effectively at an emergency scene.

_____ 6. Thirst is a reliable indicator of dehydration.

_____ 7. On a cold day, coffee or hot chocolate are appropriate beverages to rehabilitate fire fighters.

_____ 8. Rest should begin as soon as you arrive for rehabilitation.

_____ 9. Carbohydrates are used by the body to grow and repair tissues and are used as a primary fuel source only in extreme conditions such as starvation.

_____ 10. The concept of rehabilitation needs to be addressed at all types of incidents.

Short Answer

Complete this section with short written answers using the space provided.

1. Describe the types of meals a fire fighter should eat during rehabilitation for short and extended incidents.

2. Identify the personal responsibilities of each fire fighter during rehabilitation.

3. Why are caffeinated and sugar-rich beverages not recommended for rehabilitation?

4. Why is thirst not a good indicator of dehydration?

5. Which physical and mental symptoms may develop if the fire fighter does not get the opportunity to rest and recover?

6. What are the possible effects of dehydration on the fire fighter?

7. In addition to structure fires, what are some other incidents where full rehabilitation stations may be required?

8. Identify the seven functions of rehabilitation.

9. Identify some of the factors that cause firefighting to be a stressful work environment.

Fundamentals of Fire Fighter Skills

CLUES

Across

4 To restore someone or something to a condition of health or to a state of useful and constructive activity.

5 The source of energy for the body. One of the basic sugars, it is the body's primary fuel, along with oxygen.

6 A localized condition that occurs when the layers of the skin and deeper tissue freeze. (NFPA 704)

Down

1 Certain salts and other chemicals that are dissolved in body fluids and cells. Proper levels need to be maintained for good health and strength.

2 A condition in which the internal body temperature falls below 95°F (35°C), usually a result of prolonged exposure to cold or freezing temperatures.

3 A state in which fluid losses are greater than fluid intake into the body. If left untreated, this condition may lead to shock and even death.

Word Fun

The following crossword puzzle is an activity provided to reinforce correct spelling and understanding of terminology associated with firefighting. Use the clues provided to complete the puzzle. Do not include spaces or punctuation when filling in the puzzle.

Fire Alarms

The following real case scenarios will give you an opportunity to explore the concerns associated with fire fighter rehabilitation. Read each scenario, and then answer each question in detail.

1. After fighting a winter-time (temperature 15°F) defensive fire for 2 hours, your partner comments that he has been wet for the last hour, and his feet and hands are hurting and causing a great deal of pain. How should you help him?

2. You have just finished clearing brush and shrubs from the path of a wildland fire. It is a warm, 85°F summer day with no breeze. You begin to feel warm, and you are exhausted, sweating profusely, and feeling lightheaded. What should you do?

Fire Fighter II in Action

The following scenario will give you an opportunity to apply your firefighting knowledge and your fire department SOGs to the new information you learned while studying this chapter. Research your department's SOGs and answer the assignment in detail. Compare your answers with your classmates' and discuss similarities and obvious differences between your answers.

Most fire fighters work in an area where the weather and temperatures can go to extremes. In the south, temperatures can reach into the 100°F (37.8°C) range for days and weeks at a time. In the north, temperatures may also exceed 100°F (37.8°C), and may also drop below freezing for long periods of time.

1. What should your company carry on the rig to aid in rehabilitation, and also give some relief to the fire fighters during incidents in these extreme temperatures?

2. What should you carry in a personal bag or tote that will help you or your partner to cope with extended exposure to heat or cold?

Wildland and Ground Fires

Workbook Activities

The following activities have been designed to help you. Your instructor may require you to complete some or all of these activities as a regular part of your fire fighter training program. You are encouraged to complete any activity that your instructor does not assign as a way to enhance your learning in the classroom.

Chapter Review

The following exercises provide an opportunity to refresh your knowledge of this chapter.

Matching
Match each of the terms in the left column to the appropriate definition in the right column.

_____ 1. Slash **A.** Fuel that ignites and burns easily

_____ 2. Heel of the fire **B.** A new fire that starts outside the perimeter of the main fire

_____ 3. Head of the fire **C.** An area that has already been burned

_____ 4. Spot fire **D.** The traveling edge of a fire

_____ 5. Black **E.** The leftovers of a logging and land-clearing operation

_____ 6. Pocket **F.** A planned operation to remove fuel by burning out large selected areas

_____ 7. Island **G.** The amount of fuel present in a given area

_____ 8. Backfiring **H.** An unburned area surrounded by burned land

_____ 9. Wildland **I.** The side opposite the head of the fire

_____ 10. Fine fuel **J.** Land in an uncultivated natural state that is covered by timber, woodland, brush, or grass

_____ 11. Fuel volume **K.** The unburned area between a finger and the main body of the fire

Multiple Choice
Read each item carefully, and then select the best response.

_____ 1. The changes of elevation in the land as well as the positions of natural and human-made features is
  A. geography.
  B. geology.
  C. physiology.
  D. topography.

_____ 2. A firefighting attack that involves building a fire line along natural fuel breaks, along favorable breaks in topography, or at considerable distance from the fire and burning out the intervening fuel is called a(n)
  A. mounted attack.
  B. indirect attack.
  C. direct attack.
  D. counter attack.

CHAPTER 21

_____ 3. The technique used to remove fuel by burning is called
  A. adze.
  B. backfiring.
  C. direct attack.
  D. flanking.

_____ 4. A firefighting attack that requires only one team of fire fighters is called a(n)
  A. pincer attack.
  B. backfiring attack.
  C. flanking attack.
  D. indirect attack.

_____ 5. How much water do small apparatus used for fighting wildland fires typically carry?
  A. 800 gallons (3028 liters)
  B. 200–300 gallons (757–1136 liters)
  C. 2000 gallons (7571 liters)
  D. 50–100 gallons (189–379 liters)

_____ 6. Which combination tool is used to create a fire line?
  A. McLeod fire tool
  B. Adze
  C. Pulaski axe
  D. Halligan tool

_____ 7. What is the top priority in a wildland fire attack?
  A. Containment
  B. Extinguishment
  C. Minimization of damage
  D. Safety

_____ 8. A firefighting attack that requires two teams of fire fighters attacking both flanks of a wildland fire is called a(n)
  A. pincer attack.
  B. backfiring attack.
  C. flanking attack.
  D. counter attack.

_____ 9. Which term describes the relative closeness of wildland fuels?
  A. Fuel compactness
  B. Fuel continuity
  C. Fuel volume
  D. Fuel moisture

_____ 10. The firefighting attack most often used for large wildland and ground fires that are too dangerous for a direct attack is the
  A. pincer attack.
  B. backfiring attack.
  C. flanking attack.
  D. indirect attack.

___ 11. An unburned area between a finger and the traveling (main body) edge of the fire is called a(n)
   A. island.
   B. lapse.
   C. pocket.
   D. spot fire.

___ 12. Unplanned and uncontrolled fires burning in vegetative fuels that sometimes include structures are called
   A. ground cover fires.
   B. aerial fires.
   C. wildland fires.
   D. urban fires.

___ 13. As wildland and ground fires grow and reach into areas with new fuel, the traveling edge of the fire is called the
   A. heel of the fire.
   B. head of the fire.
   C. rear of the fire.
   D. arm of the fire.

___ 14. The partly decomposed organic material on a forest floor is called
   A. ground duff.
   B. slash.
   C. medium fuel.
   D. heavy fuel.

___ 15. The three causes of wildland fires are natural, accidental, and
   A. intentional fires.
   B. occupational fires.
   C. combustion fires.
   D. mechanical fires.

___ 16. Fuels that are located close to the surface of the ground are considered
   A. aerial fuels.
   B. subsurface fuels.
   C. supersurface fuels.
   D. surface fuels.

Vocabulary

Define the following terms using the space provided.

1. Heavy fuels:

   _____
   _____
   _____

2. Fuel continuity:

   _____
   _____
   _____

3. Backpack pump extinguisher:

___

4. Topography:

___

5. Aerial fuels:

___

Fill-in
Read each item carefully, and then complete the statement by filling in the missing word(s).

1. For small fires with a light fuel load, _____ _____ _____ may be an effective firefighting tactic.

2. The relative _____ is the ratio of the amount of water vapor present in the air compared to the maximum amount the air can hold at a given temperature.

3. The second side of the fire triangle is _____.

4. The location where a wildland or ground fire begins is called the _____ _____ _____.

5. Vegetative fuels can be located _____, _____, or _____ the ground.

6. A(n) _____ fire is a new fire that starts outside the perimeter of the main fire.

7. _____-wing aircraft can take on a load of water from a lake and apply it to the fire.

8. _____ conditions have a major impact on the behavior of wildland fires.

9. _____ and _____ fires can advance and change directions quickly.

10. Fires spread more _____ in fine fuels than in heavy timber and brush.

True/False

If you believe the statement to be more true than false, write the letter "T" in the space provided. If you believe the statement to be more false than true, write the letter "F."

_____ 1. The two most critical weather conditions that influence a wildland fire are moisture and wind.

_____ 2. Fire shelters can be carried in a protective pouch on a fire fighter's belt.

_____ 3. The amount of moisture in a fuel is related to the season of the year.

_____ 4. A direct attack on a wildland fire is made by attacking the left flank of the main body.

_____ 5. Roots, moss, duff, and decomposed stumps are examples of heavy fuels.

_____ 6. Rising of heated air in a wildland fire will preheat the fuels above the main body of the fire.

_____ 7. The fire triangle consists of three elements: fuel, oxygen, and heat.

_____ 8. Wildland fires are unplanned and uncontrolled fires burning in vegetative fuel that sometimes includes structures.

_____ 9. Fine fuels have a small surface area relative to their volume.

_____ 10. When relative humidity is high, the moisture from the air is absorbed by vegetative fuels, making them less susceptible to ignition.

Short Answer

Complete this section with a short written answer using the space provided.

1. List three hazards of wildland fires.

_____
_____
_____
_____
_____

Word Fun

The following crossword puzzle is an activity provided to reinforce correct spelling and understanding of terminology associated with firefighting. Use the clues provided to complete the puzzle. Do not include spaces or punctuation when filling in the puzzle.

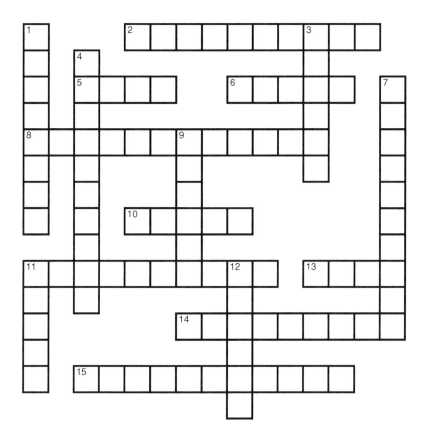

CLUES

Across

2. The land surface configuration.
5. A system that combines foam concentrate, water, and compressed air to produce foam that can stick to both vegetation and structures.
6. An area that has already been burned.
8. The amount of moisture present in a fuel, which affects how readily the fuel will ignite and burn.
10. Debris resulting from natural events such as wind, fire, snow, or ice breakage, or from human activities such as building or road construction, logging, pruning, thinning, or brush cutting.
11. Partly decomposed organic material on a forest floor; a type of light fuel.
13. A hand tool constructed of a thin, arched blade set at right angles to the handle. It is used to chop brush for clearing a fire line, or to mop up a wildland fire.
14. Fuels that ignite and burn easily, such as dried twigs, leaves, needles, grass, moss, and light brush.
15. An item of protective equipment configured as an aluminized tent utilized for protection, by means of reflecting radiant heat, in a fire entrapment situation.

Down

1. A new fire that starts outside areas of the main fire, usually caused by flying embers and sparks.
3. A deep indentation of unburned fuel along the fire's perimeter, often found between a finger and the head of the fire.
4. A hand tool used for constructing fire lines and overhauling wildland fires. One side of the head consists of a five-toothed to seven-toothed fire rake; the other side is a hoe.
7. Fuels of a large diameter, such as large brush, heavy timber, snags, stumps, branches, and dead timber on the ground. These fuels ignite and are consumed more slowly than light fuels.
9. An unburned area surrounded by fire.
11. An area of unburned fuels.
12. A narrow point of fire whose extension is created by a shift in wind or a change in topography.

Fire Alarms

The following real case scenarios will give you an opportunity to explore the concerns associated with wildland and ground fires. Read each scenario, and then answer each question in detail.

1. You are clearing brush and extinguishing small fires in your assigned area when suddenly the wind picks up and shifts, building the fire and driving it toward you. Your escape route has been cut off and the flames are approaching. How will you protect yourself?

   _____
   _____
   _____
   _____
   _____
   _____

2. You are assigned to drive the brush truck for the first time. It is a larger-sized truck with a water capacity of 1000 gallons (3785 liters). You are dispatched to a wildland fire in a hilly, undeveloped part of your response district. As you approach the fire, you are instructed to drive down a very steep, unimproved road. How will you proceed?

   _____
   _____
   _____
   _____
   _____
   _____

Fire Fighter II in Action

The following scenario will give you an opportunity to apply your firefighting knowledge and your fire department SOGs to the new information you learned while studying this chapter. Research your department's SOGs and answer the assignment in detail. Compare your answers with your classmates' and discuss similarities and obvious differences between your answers.

Your community has just acquired several hundred acres of undeveloped property, most of it hilly and forested. Your department is inexperienced at wildland firefighting, and has no wildland firefighting personal protective equipment (PPE) or equipment. Your Chief has assured all of you the training and equipment will be provided, but city council does not see the need for the equipment, and is not releasing the necessary funds.

1. What are your concerns with using your structural firefighting PPE during grass and wildland firefighting?

   _____
   _____
   _____
   _____
   _____

2. What, if any, additional training do you believe should be provided to your department?

   _____
   _____
   _____
   _____
   _____

3. What additional equipment would make fighting fires in this area safer for the fire fighters?

___

Skill Drills

Skill Drill 21-2: Suppressing a Ground Cover Fire
Test your knowledge of the skill drill by filling in the correct words in the photo captions.

© Jones & Bartlett Learning. Photographed by Glen E. Ellman.

1. Don appropriate PPE. Identify _____ and _____ risks. Protect exposures if necessary.

© Jones & Bartlett Learning. Photographed by Glen E. Ellman.

2. Construct a fire line by removing _____ with hand tools. OR

© Jones & Bartlett Learning. Photographed by Glen E. Ellman.

3. Extinguish the fire with a(n) _____ pump extinguisher or a hand line.

© Jones & Bartlett Learning. Photographed by Glen E. Ellman.

4. _____ the area completely to ensure complete extinguishment of the ground cover.

Fire Suppression

Workbook Activities

The following activities have been designed to help you. Your instructor may require you to complete some or all of these activities as a regular part of your fire fighter training program. You are encouraged to complete any activity that your instructor does not assign as a way to enhance your learning in the classroom.

Chapter Review

The following exercises provide an opportunity to refresh your knowledge of this chapter.

Matching

Match each of the terms in the left column to the appropriate definition in the right column.

_____ 1. Soffit
_____ 2. Master stream device
_____ 3. Indirect attack
_____ 4. Combination attack
_____ 5. Elevated master stream device
_____ 6. Direct attack
_____ 7. Portable monitor
_____ 8. Three team members
_____ 9. Deck gun
_____ 10. Solid stream

A. Employs both direct and indirect attack methods sequentially
B. Apparatus-mounted device intended to flow large amounts of water directly onto a fire or exposed building
C. An elevated master stream device that is mounted at the tip of an aerial ladder
D. Produced by a smooth-bore nozzle
E. Usually needed to advance a 2½-inch hose line inside a building
F. Used to produce high-volume water streams
G. A master stream device that can be positioned wherever a master stream is needed
H. The material covering the gap between the edge of the roof and the exterior wall of the house
I. Used to remove as much heat as possible
J. A straight or solid hose stream directed at the base of the fire

Multiple Choice

Read each item carefully, and then select the best response.

_____ 1. When controlling and extinguishing electrical or electrical equipment fires, the fire fighter uses
   A. Class A agents.
   B. Class B agents.
   C. Class C agents.
   D. Class D agents.

CHAPTER 22

_____ 2. Master stream devices are used
   A. to produce high-volume water streams for large fires.
   B. to produce low-volume water streams for small fires.
   C. to produce increased air movement and ventilation.
   D. because they give the greatest control over the water flow.

_____ 3. When actions are taken to prevent the spread of a fire to areas that are not already burning, it is called
   A. hosing down.
   B. maximum coverage.
   C. master stream spray.
   D. protecting exposures.

_____ 4. Master streams are typically used for
   A. structure fires.
   B. hazardous materials incidents.
   C. offensive operations.
   D. defensive operations.

_____ 5. What is the primary objective in a defensive operation?
   A. To ensure the least amount of property damage
   B. To provide a safe environment for the fire fighter
   C. To prevent the fire from spreading
   D. To prepare the fire fighter for offensive attacks

_____ 6. To decrease the need for manual overhaul during trash container fires, it is very useful to apply
   A. Class A foam.
   B. Class B foam.
   C. Class C foam.
   D. Class D foam.

_____ 7. An offensive fire attack initiated by an exterior, indirect handline operation is a
   A. master stream attack.
   B. cellar nozzle attack.
   C. transition attack.
   D. deluge attack.

_____ 8. An initial exterior attack can be started using a preconnected hose line or a
   A. deck gun.
   B. penetrating nozzle.
   C. Class C extinguisher.
   D. purple K extinguisher.

_____ 9. During vehicle fires under the hood or engine area, fire fighters should approach from the
   A. downhill and downwind side.
   B. downhill and upwind side.
   C. uphill and downwind side.
   D. uphill and upwind side.

_____ 10. Flammable liquid fires can be extinguished using
   A. Class A foam.
   B. Class B foam.
   C. Class C foam.
   D. Class D foam.

_____ 11. The pattern and form of the water that is discharged onto the fire is defined by the
   A. nozzle.
   B. hose line.
   C. fire size.
   D. strategy involved.

_____ 12. After the main body of a vehicle fire has been extinguished, it is important to
   A. continue to apply a master stream from a safe distance.
   B. remove leftover flammable liquids.
   C. chock the wheels.
   D. contain water overflow.

_____ 13. The decision on what type of fire attack to implement is made by the
   A. fire fighter.
   B. captain.
   C. incident commander.
   D. fire chief.

_____ 14. When fire fighters begin with an indirect attack and then continue with a direct attack, which type of attack are they utilizing?
   A. Aggressive
   B. Combination
   C. Multiple
   D. Progressive

_____ 15. Large handlines are defined as hoses that have a diameter of at least
   A. 2½ inches.
   B. 3 inches.
   C. 3½ inches.
   D. 5 inches.

_____ 16. When water is converted to steam, it expands to occupy a volume that is
   A. 10,000 times greater than the volume of water.
   B. 7000 times greater than the volume of water.
   C. 1700 times greater than the volume of water.
   D. 1000 times greater than the volume of water.

_____ 17. In situations where the temperature is increasing and it appears that the room or space is ready to experience flashover, fire fighters should use a(n)
   A. indirect attack.
   B. exterior attack.
   C. indirect application of water.
   D. direct application of water.

_____ 18. What type of fire attack, when successful, controls the fire with the least amount of property damage?
   A. Offensive
   B. Defensive
   C. Transition
   D. Combination

_____ 19. Vehicles that use compressed natural gas are powered by cylinders usually located
   A. under the hood of the vehicle.
   B. in the trunk of the vehicle.
   C. under the chassis of the vehicle.
   D. in front of the driver-side engine compartment.

____ **20.** To prevent explosions in possible overheating situations, propane tanks are equipped with
   A. relief valves.
   B. release valves.
   C. connection valves.
   D. vapor space.

____ **21.** Storing propane as a liquid is very efficient because it has an expansion ratio of
   A. 27:1.
   B. 170:1.
   C. 270:1.
   D. 370:1.

Vocabulary

Define the following terms using the space provided.

1. Master stream device:

2. Indirect application of water:

3. Portable monitor:

4. Straight stream:

5. Boiling-liquid, expanding-vapor explosion (BLEVE):

Fundamentals of Fire Fighter Skills

Fill-in
Read each item carefully, and then complete the statement by filling in the missing word(s).

1. It is more difficult for fire fighters to advance and maneuver a(n) _____ handline inside a building.

2. A(n) _____ stream divides water into droplets, which have a very large surface area and can absorb heat efficiently.

3. Directing water onto a fire from a safe distance is a(n) _____ operation.

4. If fire fighters do not identify the basement fire prior to entering the building above the basement fire, they are at an _____ _____ for falling through the damaged floor and ending up in the burning basement below.

5. Wetting neighboring exposures will keep the fuel from reaching its _____ temperature.

6. A fog pattern used in an _____ _____ _____ will entrain large volumes of air and pushes air into the building.

7. Large handlines and master streams are more often used in _____ operations.

8. An initial exterior attack can be started using water from the _____ _____ on the apparatus even before a permanent water supply has been established.

9. A _____ _____ is an offensive fire attack initiated by an exterior, indirect handline operation into the fire compartment to initiate cooling while transitioning into interior direct fire attack in coordination with ventilation operations.

10. Fire fighters who enter a confined space should carry a _____ device.

11. A(n) _____ attack uses a straight or solid hose stream to deliver water onto the base of the fire.

True/False
If you believe the statement to be more true than false, write the letter "T" in the space provided. If you believe the statement to be more false than true, write the letter "F."

_____ 1. Master stream devices commonly flow between 350 and 2000 gpm (1591 and 9092 lpm).

_____ 2. Fires in stacked or piled materials can be approached aggressively, because they often burn evenly.

_____ 3. A master stream device can be directed by remote control.

_____ 4. The air movement created by a fog stream can be used for ventilation.

_____ 5. A portable monitor is attached to a handline to create a fog stream.

_____ 6. During vehicle fires in hybrid automobiles, the cables that connect the batteries to the electric motors must be cut as quickly as possible to prevent sparking.

_____ 7. One of the most crucial decisions made by the IC is whether to initiate a defensive or offensive fire attack.

_____ 8. Successful offensive attacks often result in the least amount of property damage.

_____ 9. The movement of fire is dependent on the flow path of heated smoke and gases.

_____ 10. A solid stream has more reach and penetrating power than a fog or straight stream.

_____ 11. The first step in an interior fire attack is to don full personal protective equipment (PPE), including a self-contained breathing apparatus (SCBA).

_____ 12. The best method to prevent a BLEVE is to direct heavy water streams onto the tank from a safe distance.

_____ 13. A transitional attack is an offensive fire attack that occurs just prior to entry, search, and tactical ventilation.

Short Answer

Complete this section with short written answers using the space provided.

1. Describe the characteristics of the following:

   A. Fog stream

   B. Straight

   C. Solid stream

2. Describe the objectives of the following:

   A. Direct attack

   B. Indirect attack

C. Combination attack

3. Describe the characteristics of concealed space fires.

4. Describe the characteristics of basement fires.

5. List three important points to remember regarding basement fires.

6. Describe the tactics used to suppress fires above ground level.

Chapter 22: Fire Suppression

7. Identify five factors to be evaluated when considering whether to enter an involved structure or to mount an attack.

8. Describe the characteristics of flammable-gas cylinders.

9. Describe the hazards presented by flammable-gas fires.

10. Describe a boiling-liquid, expanding-vapor explosion.

11. Describe tactics used to suppress flammable gas fires.

Fire Alarms

The following real case scenarios will give you an opportunity to explore the concerns associated with fire suppression. Read each scenario, and then answer each question in detail.

1. It is 9:00 in the morning when you are dispatched to a residential structure fire. While you are en route, dispatch tells you that all of the occupants have evacuated. Upon arrival, your Captain gives a size-up: a one-story, wood-frame residential structure with fire and smoke coming out of one room. He tells you and your partner to complete an offensive direct attack on the fire room. How should you proceed?

2. It is 6:30 in the evening when a thunder and windstorm hits your community. Winds are gusting up to 60 miles per hour (97 kilometers per hour). Your engine is dispatched to an electrical pole that is on fire. Upon arrival, you see that a transformer is on fire. How should you proceed?

Fire Fighter II in Action

The following scenario will give you an opportunity to apply your firefighting knowledge and your fire department SOGs to the new information you learned while studying this chapter. Research your department's SOGs and answer the assignments in detail. Compare your answers with your classmates' and discuss similarities and obvious differences between your answers.

You and your partner are advancing a 1 ¾-inch handline down a stairway into the basement of a newly constructed single-family dwelling. It is very hot and smoky in the stairwell, and difficult to see. As you advance into the basement you remember, from your preplanning of this housing development, that most of these structures had unfinished wood truss ceilings in the basement. The fire is well involved, and the basement is very hot.

1. As the nozzleman, what should you be considering?

2. What radio communications should you make to your company officer or the incident commander (IC)?

Skill Drill

Skill Drill 22-2: Performing a Direct Attack

Test your knowledge of this skill drill by placing the steps below in the correct order. Number the first step with a "1," the second step with a "2," and so on.

_____ Shut down the nozzle and listen.

_____ Don a face piece and activate the SCBA and personal alert safety system (PASS) device prior to entering the building.

_____ Advance the hose line from the apparatus to the entry point of the structure. Flake out excess hose in front of the door.

_____ Apply water in either a straight or solid stream onto the base of the fire until all visible flame has been extinguished.

_____ Make sure that ventilation is completed or in progress.

_____ Watch for changes in fire conditions.

_____ Signal the pump operator/driver that you are ready for water.

_____ Select the proper hose line to fight the fire based on the fire's size, location, and type.

_____ Open the nozzle to purge air from the system and make sure water is flowing.

_____ Exit the fire apparatus wearing full PPE, including SCBA.

_____ Enter into the structure and locate the seat of the fire.

_____ Locate and extinguish hot spots.

Skill Drill 22-3: Performing an Indirect Attack

Test your knowledge of this skill drill by placing the steps below in the correct order. Number the first step with a "1," the second step with a "2," and so on.

_____ Select the correct hose line to be used to attack the fire depending on the type of fire, its location, and its size.

_____ Attack any remaining fire and hot spots until the fire is completely extinguished.

_____ Watch for changes and a reduction in the amount of fire. Once the fire is reduced, shut down the nozzle.

_____ Exit the fire apparatus wearing full PPE, including SCBA.

_____ Advance the hose line from the apparatus to the opening in the structure where the indirect attack will be made.

_____ Confirm that ventilation has been completed.

_____ Advance with a charged hose line to the location where you will apply water.

_____ Don a face piece, and activate the SCBA and PASS device.

_____ Open the nozzle to make sure that air is purged from the hose line and that water is flowing. If using a fog nozzle, ensure that it is set to the proper nozzle pattern for entry. Shut down the nozzle until you are in a position to apply water.

_____ Direct the water stream toward the upper levels of the room and ceiling into the heated area overhead, and move the stream back and forth. Flow water until the room begins to darken. Shut the nozzle off, and reassess the fire conditions.

_____ Notify the operator/driver that you are ready for water.

Skill Drill 22-4: Performing a Combination Attack

Test your knowledge of this skill drill by placing the steps below in the correct order. Number the first step with a "1," the second step with a "2," and so on.

_____ Aim the nozzle at the upper-left corner of the fire and make either a "T," "O," or "Z" pattern with the nozzle. Start high and then work the pattern down to the fire level.

_____ Don full PPE and SCBA. Select the correct hose line to accomplish the suppression task at hand.

_____ Enter the structure, and locate the room or area where the fire originated.

_____ Open the nozzle to get the air out and make sure that water is flowing.

_____ Stretch the hose line to the entry point of the structure, and signal the operator/driver that you are ready to receive water.

_____ Use only enough water to darken down the fire without upsetting the thermal layering.

_____ Once the fire has been reduced, find the remaining hot spots and complete fire extinguishment using a direct attack.

Fundamentals of Fire Fighter Skills

Skill Drill 22-5: Performing the One-Fire-Fighter Method for Operating a Large Handline
Test your knowledge of the skill drill by filling in the correct words in the photo captions.

© Jones & Bartlett Learning. Photographed by Glen E. Ellman.

1. Select the correct size of fire hose. Advance the hose into position. Signal that you are ready for water and open the nozzle to allow air to escape and to ensure that water is flowing. Close the nozzle and then make a loop with the hose, ensuring that the nozzle is _____ the hose line that is coming from the fire apparatus.

© Jones & Bartlett Learning. Photographed by Glen E. Ellman.

2. Lash the hose sections together where they cross, or use your body weight to kneel or sit on the hose line at the point where the hose _____ _____.

© Jones & Bartlett Learning. Photographed by Glen E. Ellman.

3. Allow enough hose to extend past the section where the line crosses itself for _____.

© Jones & Bartlett Learning. Photographed by Glen E. Ellman.

4. Open the nozzle and _____ water onto the designated area.

Skill Drill 22-6: Performing the Two-Fire-Fighter Method for Operating a Large Handline

Test your knowledge of the skill drill by filling in the correct words in the photo captions.

1. _____ the hose line from the fire apparatus into position.

2. _____ that you are ready for water and open the nozzle to allow air to escape and to ensure water is flowing. Advance the hose line as needed.

3. Before attacking the fire, the fire fighter on the nozzle should cradle the hose on his or her hip while grasping the nozzle with one hand and supporting the hose with the other hand. The second fire fighter should stay approximately _____ feet behind the fire fighter who is on the nozzle. The second fire fighter should grasp the hose with two hands and may use a knee to stabilize the hose against the ground if necessary.

4. Open the _____ in a controlled fashion and direct water onto the fire or designated exposure.

Preincident Planning

Workbook Activities

The following activities have been designed to help you. Your instructor may require you to complete some or all of these activities as a regular part of your fire fighter training program. You are encouraged to complete any activity that your instructor does not assign as a way to enhance your learning in the classroom.

Chapter Review

The following exercises provide an opportunity to refresh your knowledge of this chapter. All questions in this chapter are Fire Fighter II level.

Matching

Match each of the terms in the left column to the appropriate definition in the right column.

_____ 1. Drafting sites

_____ 2. Target hazard

_____ 3. Fire load

_____ 4. Static water supply

_____ 5. Tanker shuttle

_____ 6. Standpipe system

_____ 7. Conflagration

_____ 8. Horizontal evacuation

_____ 9. Sprinkler system

_____ 10. HVAC

_____ 11. Defend-in-place

_____ 12. Exposure

_____ 13. Lightweight construction

_____ 14. Preincident planning

_____ 15. Size-up

A. Ongoing observation and evaluation of factors that influence the objectives, strategy, and tactics for fire suppression

B. System of tankers that transports water from a water source to the fire scene

C. Arrangement of piping, valves, and hose connections installed in a structure to deliver water for fire hoses on each floor of a building

D. Heating, ventilation, and air-conditioning system

E. The process of obtaining information about a building or a property and storing the information in a system so that it can be retrieved quickly for future reference

F. The amount of combustible material and the rate of heat release

G. Where a pumper can draft water directly from a static source

H. An integrated system of underground and overhead piping designed in accordance with fire protection engineering standards. The installation includes at least one automatic water supply that supplies one or more systems.

I. Properties that pose an increased risk to fire fighters

J. A strategy in which victims are protected from the fire without relocation

K. A large fire, often involving multiple structures

L. Property that may be endangered by flames, smoke, gases, heat, or runoff from a fire

M. A lake or a stream

N. Uses assemblies of small components, such as trusses or fabricated beams, as structural support materials

O. A strategy of moving occupants from a dangerous area to a safe area on the same floor level

CHAPTER 23

Multiple Choice

Read each item carefully, and then select the best response.

_____ 1. The preincident survey should be conducted in
   A. a systematic, uniform format.
   B. normal duty uniform.
   C. scheduled annual visits.
   D. teams of four.

_____ 2. The most challenging problem during an emergency incident at a healthcare facility
   A. is the limited access to patients.
   B. is the limitations of the floor plans.
   C. is protecting nonambulatory patients.
   D. is negotiating traffic en route to the facility.

_____ 3. Lightweight construction
   A. can be found only in newer buildings.
   B. utilizes trusses as structural support materials.
   C. is the sturdiest of the newer construction types.
   D. is the most cost-effective type of construction.

_____ 4. The classifications of buildings by major use group are
   A. lightweight, tested development, heavyweight, and open.
   B. public assembly, institutional, commercial, and industrial.
   C. renovated private, renovated public, commercial development, and industrial.
   D. public assembly, institutional, and commercial.

_____ 5. Facility security and personnel safety are both major concerns at
   A. schools and daycare centers.
   B. hospitals and nursing homes.
   C. residential occupancies.
   D. detention and correctional facilities.

_____ 6. Drafting sites should be included in a preincident plan because they identify
   A. which adjoining structures are most susceptible to fire spread.
   B. open areas that can trap fire fighters.
   C. the best routes for ventilation.
   D. locations where an engine can draft water directly from a static source.

_____ 7. A preincident plan should include information about
   A. a building's floor plan.
   B. entrance and exit locations.
   C. hazardous materials stored in the building.
   D. all of the above.

_____ 8. The primary role of a fire alarm system is to
   A. alert the occupants of a building when an incident occurs.
   B. alert the fire department of an incident.
   C. meet safety standards of the building code.
   D. all of the above.

246 FUNDAMENTALS OF FIRE FIGHTER SKILLS

_____ 9. The five types of building construction in descending order of fire resistance are
   A. fire resistive, noncombustible, ordinary, heavy timber, and wood frame.
   B. ordinary, noncombustible, fire resistive, wood frame, and heavy timber.
   C. fire resistive, noncombustible, ordinary, wood timber, and heavy frame.
   D. ordinary, fire resistive, noncombustible, wood frame, and heavy timber.

_____ 10. Horizontal ventilation can be accessed through
   A. windows and doors.
   B. windows and chimneys.
   C. ceiling and pressure fans.
   D. windows and skylights.

Vocabulary

Define the following terms using the space provided.

1. Dry hydrant:
   _____
   _____
   _____

2. Conflagration:
   _____
   _____
   _____

3. Preincident plan:
   _____
   _____
   _____

4. Fire alarm annunciator panel:
   _____
   _____
   _____

5. Heating, ventilation, and air-conditioning (HVAC) system:
   _____
   _____
   _____

6. Ordinary construction:

_____
_____
_____

7. Material Safety Data Sheet (MSDS) document:

_____
_____
_____

Fill-in
Read each item carefully, and then complete the statement by filling in the missing word(s).

1. Commercial use classification structures include the occupancy subcategories of _____ _____, _____, _____ _____, parking garages, and warehouses.

2. During an emergency situation, it may be necessary to shut off the utilities such as _____ or _____ _____ as a safety measure.

3. A high-rise building is generally defined as a structure that is more than _____ feet high.

4. If the quantity of hazardous materials on hand exceeds a specified limit, federal and state regulations require a property owner to provide the local fire department with current inventories and a(n) _____ _____ _____ _____ (_____).

5. _____ _____ are installed in high-rise buildings to eliminate the need to extend hose lines from a pumper at the street level up to the fire level.

6. The _____-_____-_____ philosophy presumes that patients or occupants of some facilities will not be able to escape from a fire without assistance and, therefore, the facility itself is designed to protect patients from the fire.

7. Moving patients from a dangerous area to a safer area on the same floor is known as _____ evacuation.

8. When the dimensions of the interior materials are greater than the dimensions of ordinary construction, the building construction is considered Type _____ or _____ _____.

9. Schools and hospitals are in the _____ major use classification.

10. In preparation for possible search-and-rescue operations, it will greatly assist the fire fighters' efforts if they know where the _____ and _____ are located.

11. The preincident survey should consider both _____ and _____ access problems.

12. If the water is obtained from a lake or a stream it is considered a(n) _____ _____ _____.

13. Wood-frame buildings are classified as Type _____.

14. A Type I building can also be referred to as fire resistive and includes materials such as _____, _____ _____, and _____ _____ _____.

15. A Type II building can also be referred to as _____ and is made of structure members that are noncombustible materials, but may not have _____ _____ _____.

16. A(n) _____ is any other building or item that may be in danger if an incident occurs in another building or area.

17. _____-_____ is the ongoing observation and evaluation of factors that are used to develop objectives, strategy, and tactics for fire suppression.

18. Preincident plans should include the most efficient route to a property and a(n) _____ route in case of traffic interruptions.

19. The process of obtaining information about a building and storing the information in a system so that it can be retrieved quickly for future reference is often referred to as _____ _____.

20. A preincident _____ is used by a team to gather information about a property to develop a preincident plan.

21. The _____ _____ can use the preincident information to direct the emergency operations more effectively.

22. A properly designed and maintained automatic _____ _____ can help control or extinguish a fire before the arrival of the fire department.

23. The _____ _____ is the part of the fire alarm system that indicates the location of an alarm within a building.

24. The use of _____ _____ _____ has greatly increased the ability of fire departments to capture, store, organize, update, and quickly retrieve preincident planning information.

25. Building layout and access information is particularly important during the _____ phase of an emergency incident.

26. Properties that pose an increased risk to fire fighters during an emergency response are identified as _____ _____.

True/False

If you believe the statement to be more true than false, write the letter "T" in the space provided. If you believe the statement to be more false than true, write the letter "F."

_____ 1. The preincident survey is an excellent time to identify the best locations for placing ground ladders or using aerial apparatus.

_____ 2. Buildings with unprotected steel beams are Type I: Fire Resistive, according to NFPA 220, *Standard on Types of Building Construction*.

_____ 3. Lightweight construction uses materials too light to cause injury to a fire fighter in proper personal protective equipment.

_____ 4. All properties have the potential to create a conflagration.

_____ 5. The objective of the preincident plan is to provide information for more effective operations during emergency incidents.

_____ 6. Potential natural barricades should be included in the preincident plan.

_____ 7. Preincident surveys of commercial and industrial properties should be conducted by independent contractors through the fire department.

_____ 8. Wood-frame building construction has floors and walls made of combustible wood material.

_____ 9. A target hazard is any property large enough to catch fire.

_____ 10. The fire load is the amount of combustible material and the rate of heat release a property may include.

Short Answer

Complete this section with short written answers using the space provided.

1. Describe why and for which types of properties a preincident plan is created.

   _____
   _____
   _____
   _____
   _____
   _____
   _____

2. List typical target hazard properties that may be found in a community.

   _____
   _____
   _____
   _____
   _____
   _____
   _____

3. Describe how a preincident survey is performed.

   _____
   _____
   _____
   _____
   _____
   _____

Fundamentals of Fire Fighter Skills

4. List the information that is gathered during a preincident survey.

_____
_____
_____
_____
_____
_____

5. Describe the information that needs to be gathered to assist the incident commander in making a rapid and correct size-up during an emergency incident.

_____
_____
_____
_____
_____
_____

6. Explain how to identify built-in fire detection and suppression systems during a preincident survey.

_____
_____
_____
_____
_____
_____

7. Describe the tactical information that is collected during a preincident survey.

_____
_____
_____
_____
_____

8. Describe how the sources of water supply for fire suppression operations are identified.

_____
_____
_____
_____
_____

9. Describe how preincident planning for the following is performed:

A. Search and rescue:

B. Rapid forcible entry:

C. Safe ladder placement:

D. Effective ventilation:

10. List the occupancy considerations to take into account when conducting a preincident survey of the following:

A. High-rise buildings:

B. Assembly occupancies:

C. Healthcare facilities:

D. Detention and correctional facilities:

E. Residential occupancies:

11. List the types of locations that require special considerations in preplanning.

Word Fun

The following crossword puzzle is an activity provided to reinforce correct spelling and understanding of terminology associated with firefighting. Use the clues provided to complete the puzzle. Do not include spaces or punctuation when filling in the puzzle.

CLUES

Across

1. Any person or property that could be endangered by fire, smoke, gases, runoff, or other hazardous conditions.
4. The weight of combustibles in a fire area or on a floor in buildings and structures, including either contents or building parts, or both.
5. A written document resulting from the gathering of general and detailed information to be used by public emergency response agencies and private industry for determining the response to reasonably anticipated emergency incidents at a specific facility.
6. Type of construction in which lightweight materials or advanced engineering, or both practices, result in a weight saving without sacrifice of strength or efficiency.
10. A form, provided by manufacturers and compounders (blenders) of chemicals, containing information about chemical composition, physical and chemical properties, health and safety hazards, emergency response, and waste disposal of the material.
11. Ongoing observation and evaluation of factors that influence the objectives, strategy, and tactics for fire suppression.
12. Any occupancy type or facility that presents a high potential for loss of life or serious impact to the community resulting from fire, explosion, or chemical release.

8. Another word for a tanker.
9. A system of computer software, hardware, data, and personnel to describe information tied to a spatial location.

Down

2. For fire protection purposes, an integrated system of underground and overhead piping designed in accordance with fire protection engineering standards.
3. A large fire, often involving multiple structures.
7. A system to manage the internal environment that is often found in large buildings.

Fundamentals of Fire Fighter Skills

Fire Alarms

The following real case scenarios will give you an opportunity to explore the concerns associated with preincident planning. Read each scenario, then answer each question in detail.

1. Your monthly probationary assignment is to list the target hazard properties in your first-response district.

    A. What are typical target hazard properties that could be included in your assignment?

    _____
    _____
    _____
    _____
    _____
    _____

    B. Include a sublist of properties with increased life-safety hazards for extra credit.

    _____
    _____
    _____
    _____
    _____
    _____

2. Your company has been assigned a list of buildings that must have a preincident survey conducted. As the rookie, you have been assigned the task of developing a plan for conducting each survey. Which steps are necessary for conducting a good preincident survey?

    _____
    _____
    _____
    _____
    _____
    _____

Fire Fighter II in Action

The following scenario will give you an opportunity to apply your firefighting knowledge and your fire department SOGs to the new information you learned while studying this chapter. Research your department's SOGs and answer the assignments in detail. Compare your answers with your classmates' and discuss similarities and obvious differences between your answers.

As a fire fighter, you will be called to respond to occupancies where the dangers and hazards will not be readily apparent the first time you see these structures. You should have the advantage of being familiar with your first response district, and know what hazards you may encounter in the different structures and occupancies.

1. Identify ten businesses or occupancies in your first response district that may present you with uncommon hazards. These may be businesses that store or manufacture hazardous materials, possibly have confined spaces, open elevator shafts, hot metal baths, or acid treatment vats—the list is almost endless. You may also wish to include any new housing developments.

   _____
   _____
   _____
   _____
   _____
   _____
   _____

2. Take this list and preplan these structures. Use your department preplan SOGs and include a department fire inspector if one is available.

   _____
   _____
   _____
   _____
   _____
   _____
   _____

Fire and Emergency Medical Care

Workbook Activities

The following activities have been designed to help you. Your instructor may require you to complete some or all of these activities as a regular part of your fire fighter training program. You are encouraged to complete any activity that your instructor does not assign as a way to enhance your learning in the classroom.

Chapter Review

The following exercises provide an opportunity to refresh your knowledge of this chapter.

Matching

Match each of the terms in the left column to the appropriate definition in the right column.

_____ 1. Combination EMS system
_____ 2. Advanced life support
_____ 3. Standing orders
_____ 4. Basic life support (BLS)
_____ 5. Paramedic
_____ 6. Emergency medical responder
_____ 7. EMT
_____ 8. Advanced emergency medical technician (AEMT)
_____ 9. Medical director
_____ 10. Medical control

A. An EMS provider who has training in specific aspects of advanced life support care, including intravenous therapy, interpretation of heart rhythms, and defibrillation

B. Provided by EMS physicians who can be reached by radio or telephone during a call

C. Involves more than 110 hours of training

D. An EMT who can perform limited procedures that usually fall between those provided by an EMT and those provided by a Paramedic

E. The physician providing authorization for patient care activities in the prehospital setting

F. The fire department provides medical first response, while another agency transports the patient to the hospital emergency room

G. The individual who is likely to encounter medical emergencies as part of his or her job

H. Also called protocols, these are a type of indirect medical control; they direct EMS providers to take specific action when they encounter various types of situations

I. An EMS provider who has extensive training in advanced life support care, including intravenous therapy, pharmacology, endotracheal intubation, and other advanced assessment and treatment skills

J. Focused on rapidly evaluating a patient's condition; maintaining a patient's airway, breathing, and circulation; controlling external bleeding; preventing shock; and preventing further injury or disability by immobilizing potential spinal or other bone fractures

CHAPTER 24

Multiple Choice

Read each item carefully, and then select the best response.

_____ 1. Protecting the privacy of the people you serve is
   A. an ethical responsibility.
   B. seldom a concern.
   C. handled by the medical director.
   D. an insurance issue.

_____ 2. What are the two levels of training for basic life support providers?
   A. EMT and AEMT
   B. Fire Fighter I and EMT
   C. Fire Fighter I and EMR
   D. EMR and EMT

_____ 3. Basic life support services include
   A. administering oxygen, interpreting heart rhythms, and splinting.
   B. administering medications, oxygen, and intravenous fluids to treat shock.
   C. scene control, splinting, treating for shock, and defibrillating the heart.
   D. administering oxygen, controlling external bleeding, and lifting and moving patients.

_____ 4. The three main groups that interact with the EMS provider are
   A. fire fighters, hospital administrators, and patients.
   B. patients, public onlookers, and police officers.
   C. police officers, public onlookers, and medical directors.
   D. patients, hospital personnel, and medical directors.

_____ 5. Advanced life support services include
   A. endotracheal intubation and administration of medications.
   B. administration of intravenous fluids and interfacility transport of critical care patients.
   C. pediatric and critical care specialties.
   D. defibrillation, administration of medications, and blood diagnostics.

_____ 6. What are the levels of EMS certification, in order from lowest to highest level of training?
   A. EMR, EMT, AEMT, Paramedic
   B. EMT, Paramedic, AEMT, EMR
   C. EMR, EMT, Paramedic, AEMT
   D. EMT, AEMT, Paramedic, EMR

_____ 7. Basic life support providers can perform cardiac defibrillation by using
   A. an automated external defibrillator (AED).
   B. internal massage.
   C. medication.
   D. intravenous fluids.

_____ 8. What are the two primary types of EMS delivery systems within fire departments?
   A. Paramedic and transport
   B. EMS and transport
   C. Combination EMS and public transport
   D. Combination and fire department EMS system

FUNDAMENTALS OF FIRE FIGHTER SKILLS

_____ 9. What are the two levels of training for advanced life support providers?
   A. EMT and AEMT
   B. AEMT and EMR
   C. AEMT and Paramedic
   D. EMR and Paramedic

_____ 10. Standing orders are a type of medical control that
   A. is provided by an EMS physician who can be reached by radio during a call.
   B. directs the EMS providers to take specific actions when they encounter different situations.
   C. allows fire fighters to provide only rescue operations.
   D. is used only by the medical director.

Vocabulary

Define the following terms using the space provided.

1. Basic life support (BLS):

2. Combination EMS system:

3. Medical director:

4. Fire department EMS system:

5. Advanced life support (ALS):

6. Paramedic:

Fill-in

Read each item carefully, and then complete the statement by filling in the missing word(s).

1. In some fire departments, more than _____ percent of emergency calls are for emergency medical services.

2. _____ personnel operate as an extension of a physician and use _____ _____, _____, and/or _____ _____.

3. EMS care is offered at both _____ and _____ life support levels.

4. Some fire departments _____ _____ their personnel in both fire suppression and EMS.

5. _____ services do not include the administration of medications beyond assisting the patient with his or her prescribed medications.

6. Strict _____ _____ and effective _____ _____ efforts have been quite successful in reducing the numbers of fires, enabling the fire service to take a greater role in providing emergency medical care.

7. The _____ _____ _____ course is designed for people such as teachers and daycare providers who encounter medical emergencies as part of their jobs.

8. The mission of the fire service is to _____ _____ and protect _____.

9. _____ is the level of EMS provider that covers the causes and treatments of diseases.

10. From the caller's perspective, all calls are seen as a(n) _____. Therefore, EMS providers must always remain supportive of the patient and the caller.

True/False

If you believe the statement to be more true than false, write the letter "T" in the space provided. If you believe the statement to be more false than true, write the letter "F."

_____ 1. All EMS providers in a system must be trained to work together and coordinate their activities.

_____ 2. In most departments, the number of fire calls is much greater than the number of EMS calls.

_____ 3. In a combination EMS system, the fire department provides medical first response and another agency operates the ambulances that transport the patients.

_____ 4. There are no advantages to having EMS systems located within the fire department.

_____ 5. The National Registry of Emergency Medical Technicians registers only Paramedics.

_____ 6. CME classes are unimportant to EMS providers.

_____ 7. Most fire departments provide some level of emergency medical services, although the degree of their involvement varies.

_____ 8. Local protocols define when EMS providers should give a radio report or obtain online medical direction.

_____ 9. Local protocols are a type of indirect medical control; they direct EMS providers to take specific action when they encounter various types of situations.

Short Answer

Complete this section with short written answers using the space provided.

1. Describe appropriate interactions with patients.

2. Describe fire department EMS systems.

3. Describe a combination EMS system.

4. Identify the responsibilities of the medical director.

5. Identify the duties and abilities of an Emergency Medical Responder.

6. Identify the duties and abilities of the EMT.

7. Identify the duties and abilities of the Paramedic.

8. Describe the difference between offline (indirect) and online (direct) medical control.

Word Fun

The following crossword puzzle is an activity provided to reinforce correct spelling and understanding of terminology associated with firefighting. Use the clues provided to complete the puzzle. Do not include spaces or punctuation when filling in the puzzle.

Clues

Across

3. The first trained individual to arrive at the scene of an emergency to provide initial medical care.
5. Enacted in 1996, federal legislation that provides for criminal sanctions as well as for civil penalties for releasing a patient's protected health information in a way not authorized by the patient.
7. A physician trained in emergency medicine, designated in a leadership role for the local EMS agency. (NFPA 450)
8. An EMS provider who has training in specific aspects of Advanced Life Support care, such as intravenous therapy, interpretation of heart rhythms, and defibrillation.
9. A specific level of prehospital medical care provided by trained responders, focused on rapidly evaluating a patient's condition; maintaining a patient's airway, breathing, and circulation; controlling external bleeding; preventing shock; and preventing further injury or disability by immobilizing potential spinal or other bone fractures. (NFPA 1584)

Down

1. An EMS provider who has extensive training in Advanced Life Support care, including intravenous therapy, pharmacology, endotracheal intubation, and other advanced assessment and treatment skills.
2. The physician providing direction for patient care activities in the prehospital setting. (NFPA 473)
4. A direction or instruction for delivering patient care without online medical oversight backed by authority of the system medical director. (NFPA 450)
6. An EMS provider who has training in Basic Life Support care, including automated external defibrillation, simple airway techniques, and controlling external bleeding.
8. Functional provision of advanced airway management including intubation, advanced cardiac monitoring, manual defibrillation, establishment and maintenance of intravenous access, and drug therapy. (NFPA 1584)

Fire Alarms

The following real case scenarios will give you an opportunity to explore the concerns associated with fire care. Read each scenario, and then answer each question in detail.

1. Your department is in the process of renewing its EMS contract with the local jurisdiction. A private EMS company is also bidding on providing these services. A family member asks you to explain the benefit of having the EMS system located within the fire department. How do you respond?

   _____
   _____
   _____
   _____
   _____
   _____

2. You have just completed your probationary period as a fire fighter, and you are interested in entering the emergency medical services. Your department offers a number of EMS training programs, including EMR, EMT, and Paramedic. You ask your Lieutenant for her opinion about where to start. She suggests that you enroll in the EMT course but also encourages you to research the training requirements and decide whether you can meet the demands of the class. What can you expect in terms of the time commitment and curriculum for an EMT course?

   _____
   _____
   _____
   _____
   _____
   _____

Fire Fighter II in Action

The following scenario will give you an opportunity to apply your firefighting knowledge and your fire department SOGs to the new information you learned while studying this chapter. Research your department's SOGs and answer the assignments in detail. Compare your answers with your classmates' and discuss similarities and obvious differences between your answers.

One of your assigned morning duties at the station is to check the rescue squad for required equipment, fluid levels, and cleanliness. Several times in the past you found the squad was missing items the previous shift had used during their tour and not replaced. You are the rookie at the station, and have kept quiet and restocked the squad yourself, not wishing to cause trouble.

At 0700 this morning before you had the chance to check the squad over, your crew responded to a squad run. An elderly person had fallen on some ice and fractured a hip. The previous shift had a similar emergency the night before and failed to restock the squad with the necessary splints and back board. Your crew did not have all of the equipment needed to properly treat the patient.

1. What immediate actions would you take to assist your patient?

2. Is this failure partially your responsibility?

3. What could you have done to prevent this equipment shortfall from occurring?

4. What should you do after you return to the station?

Emergency Medical Care

Workbook Activities

The following activities have been designed to help you. Your instructor may require you to complete some or all of these activities as a regular part of your fire fighter training program. You are encouraged to complete any activity that your instructor does not assign as a way to enhance your learning in the classroom.

Chapter Review

The following exercises provide an opportunity to refresh your knowledge of this chapter.

Matching
Match each of the terms in the left column to the appropriate definition in the right column.

_____ 1. Bruise   A. An injury in which a piece of skin is left hanging by a flap
_____ 2. Dressing   B. The windpipe
_____ 3. Xiphoid process C. An irregular cut or tear through the skin
_____ 4. Sternum   D. Voice box
_____ 5. Larynx   E. The fluid part of blood
_____ 6. Laceration   F. A bandage placed directly on the wound
_____ 7. Capillaries   G. The breastbone
_____ 8. Trachea   H. The key landmark for CPR and the Heimlich maneuver
_____ 9. Avulsion   I. The tube that passes food from the throat to the stomach
_____ 10. Plasma   J. A closed wound (contusion)
_____ 11. Decapitation   K. Separation of the head from the body
_____ 12. Esophagus   L. The smallest blood vessels

Multiple Choice
Read each item carefully, and then select the best response.

_____ 1. Checking an adult for a carotid pulse should take no more than
   A. 5 seconds.
   B. 10 seconds.
   C. 20 seconds.
   D. 30 seconds.

_____ 2. If the heart cannot pump enough blood to supply the needs of the body, the victim will experience
   A. cardiogenic shock.
   B. pipe failure.
   C. anaphylactic shock.
   D. spinal shock.

CHAPTER 25

_____ 3. A "child" is defined as a person between
   A. 1 and 12 to 14 years of age.
   B. 5 and 12 years of age.
   C. 8 and 12 years of age.
   D. 8 and 18 years of age.

_____ 4. Any call that involves multiple victims is termed a
   A. mass-casualty incident.
   B. multiple-casualty incident.
   C. triage.
   D. violent incident.

_____ 5. What is the major artery in the neck?
   A. Carotid artery
   B. Radial artery
   C. Brachial artery
   D. Femoral artery

_____ 6. What is the most serious type of bleeding?
   A. Venous bleeding
   B. Arterial bleeding
   C. Capillary bleeding
   D. External bleeding

_____ 7. What is the most common cause of airway obstruction?
   A. Food
   B. Small toys
   C. Dentures
   D. The tongue

_____ 8. The concept of using protective equipment to prevent exposure to infectious diseases is known as
   A. workplace safety equipment.
   B. personal protective equipment.
   C. universal precautions.
   D. infection protection.

_____ 9. A wound where the skin stays intact is called a(n)
   A. abrasion.
   B. open wound.
   C. laceration.
   D. closed wound.

_____ 10. Oxygen is carried from the lungs to the body, and carbon dioxide back to the lungs, by the
   A. plasma.
   B. platelets.
   C. white blood cells.
   D. red blood cells.

_____ 11. The immunizations recommended for medical care providers include the tetanus vaccine and
   A. hepatitis B vaccine.
   B. hepatitis C vaccine.
   C. HIV vaccine.
   D. SARS vaccine.

_____ 12. How much blood does the average adult's circulatory system contain?
   A. Approximately 6 pints
   B. Approximately 10 pints
   C. Approximately 12 pints
   D. Approximately 15 pints

_____ 13. After a victim suffers cardiac arrest, brain damage begins within
   A. 2 to 3 minutes.
   B. 4 to 6 minutes.
   C. 7 to 9 minutes.
   D. 10 to 15 minutes.

_____ 14. What is the most effective way of expelling a foreign object that is causing airway obstruction?
   A. The Heimlich maneuver
   B. Patting or rubbing the back
   C. Coughing
   D. Throwing up

_____ 15. Extreme allergic reactions to a foreign substance can cause
   A. fluid loss.
   B. cardiogenic shock.
   C. anaphylactic shock.
   D. spinal shock.

_____ 16. During CPR on an adult, chest compressions should be at the rate of
   A. 30 compressions per minute.
   B. 60 compressions per minute.
   C. 100 compressions per minute.
   D. 120 compressions per minute.

_____ 17. The tiny air sacs in the lungs where the actual exchange of gases takes place are the
   A. alveoli.
   B. capillaries.
   C. bronchi.
   D. xiphoid.

_____ 18. The pressure wave generated by the pumping action of the heart is called the
   A. heart rate.
   B. pulse.
   C. heart rhythm.
   D. arterial push.

_____ 19. If a victim is not breathing, you must breathe for him or her. This technique is known as
   A. rescue breathing.
   B. ventilation.
   C. the ABCs.
   D. victim recovery.

_____ 20. If the airway is completely obstructed, the victim will lose consciousness in
   A. less than 1 minute.
   B. 1 to 2 minutes.
   C. 3 to 4 minutes.
   D. 5 to 6 minutes.

_____ 21. When assisting an infant's breathing, after the first two breaths, rescue breaths should follow every
   A. 2 to 4 seconds.
   B. 3 to 5 seconds.
   C. 5 to 7 seconds.
   D. 10 to 15 seconds.

_____ 22. To relieve an airway obstruction in an infant, use a combination of back slaps and
   A. the Heimlich maneuver.
   B. rescue breathing.
   C. tilts.
   D. chest thrusts.

_____ 23. What is the name of the thin flapper valve that allows air to enter the trachea, but that prevents food or water from doing so?
   A. Esophagus
   B. Epiglottis
   C. Larynx
   D. Alveoli

_____ 24. What is the most critical sign of inadequate breathing?
   A. Gasping
   B. Cyanosis
   C. Respiratory arrest
   D. Unconsciousness

_____ 25. The three critical components needed to sustain life in human beings are
   A. food, clothing, and shelter.
   B. food, water, and shelter.
   C. nutrients, clothing, and air.
   D. food, water, and oxygen.

_____ 26. A normal adult has a breathing rate of approximately
   A. 8 to 12 breaths per minute.
   B. 12 to 20 breaths per minute.
   C. 20 to 30 breaths per minute.
   D. 30 to 40 breaths per minute.

_____ 27. Brain cells begin to die if they are deprived of oxygen and nutrients for
   A. 2 to 3 minutes.
   B. 4 to 6 minutes.
   C. 7 to 9 minutes.
   D. 9 or more minutes.

_____ 28. What is another name for the voice box?
   A. Esophagus
   B. Trachea
   C. Bronchi
   D. Larynx

_____ 29. What are the smallest branches of the circulatory system, where the exchange of oxygen and carbon dioxide takes place?
   A. Veins
   B. Arteries
   C. Capillaries
   D. Blood vessels

_____ 30. Opening the airway by lifting the victim's head backward and lifting the chin forward, bringing the entire lower jaw with it, is called the
   A. jaw-thrust technique.
   B. head tilt–chin lift.
   C. head tilt and shift.
   D. jaw and head lift.

270 Fundamentals of Fire Fighter Skills

Labeling
Label the following diagram with the correct terms.

1. Anatomy of the respiratory system.

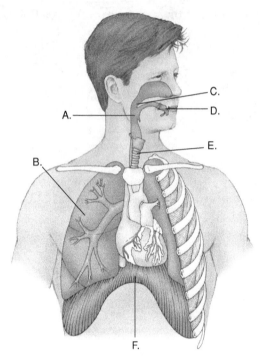

Figure 25-3

A. _____
B. _____
C. _____
D. _____
E. _____
F. _____

2. The heart.

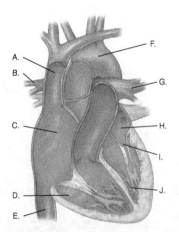

Figure 25-42

A. _____
B. _____
C. _____
D. _____
E. _____
F. _____
G. _____
H. _____
I. _____
J. _____

Vocabulary

Define the following terms using the space provided.

1. Dependent lividity:

2. Radial artery:

3. Shock:

4. Cardiopulmonary resuscitation (CPR):

5. Recovery position:

Fill-in

Read each item carefully, and then complete the statement by filling in the missing word(s).

1. The most common foreign object that causes an airway obstruction is _____.

2. Many injured victims actually die from _____ caused by blood loss.

3. The first step in assessing a victim's airway is to check the victim's level of _____.

4. If two rescuers are performing CPR on an infant, give _____ rescue breath(s) after every 15 chest compressions.

5. _____ is the virus that can lead to acquired immune deficiency syndrome.

6. If there is no pulse, you must correct the victim's circulation by performing external _____ _____.

7. The main purpose of the respiratory system is to provide _____ and to remove _____ _____ from the body.

8. The _____ rely on the diaphragm for movement, as they consist of soft, spongy tissue with no muscles.

9. To open the airway of a victim with a suspected neck injury, use the _____-_____ maneuver.

10. In a healthy individual, the oxygen saturation should be between _____ percent and _____ percent when the person is breathing room air.

11. When air is forced into the stomach instead of the lungs, _____ _____ occurs.

12. Based on the assumption that all victims are potential carriers of bloodborne pathogens, the Centers for Disease Control and Prevention (CDC) recommends all healthcare workers use _____ _____.

True/False

If you believe the statement to be more true than false, write the letter "T" in the space provided. If you believe the statement to be more false than true, write the letter "F."

_____ 1. Within 8 to 10 minutes after a cardiac arrest, the damage to the brain may be irreversible.

_____ 2. Living wills, advance directives, and DNR orders are legal documents that specify the patient's wishes regarding particular medical procedures.

_____ 3. Heavy leather gloves prevent the spread of infectious diseases.

_____ 4. Medical gloves, masks, and eye protection prevent the spread of infectious diseases.

_____ 5. Rigor mortis is an indication that a victim has been dead for more than a day.

_____ 6. The "B" in the CPR ABCs stands for "bleeding."

_____ 7. In adult CPR, after every 15 chest compressions, give 2 rescue breaths.

_____ 8. Drug-resistant strains of tuberculosis can be transmitted through the air.

_____ 9. Once a brain cell has been destroyed, it cannot be healed or replaced.

_____ 10. Mouth-to-mask devices prevent the transmission of infectious diseases.

_____ 11. Regardless of the cause of cardiac arrest, the initial treatment is the same.

_____ 12. An unconscious victim's airway is often blocked.

Short Answer

Complete this section with short written answers using the space provided.

1. List the five steps used to combat or begin treatment of shock.

2. Identify the six acceptable criteria for discontinuing CPR on a victim.

3. Identify the Centers for Disease Control and Prevention's recommended five steps for universal precautions.

4. List five of the major causes of respiratory arrest.

5. Describe blood flow through the four chambers of the heart.

_____
_____
_____
_____

6. Identify five of the possible signs and symptoms of shock.

_____
_____
_____
_____
_____

Word Fun

The following crossword puzzle is an activity provided to reinforce correct spelling and understanding of terminology associated with firefighting. Use the clues provided to complete the puzzle. Do not include spaces or punctuation when filling in the puzzle.

CLUES

Across

1. An injury in which soft-tissue damage occurs beneath the skin, even though there is no break in the surface of the skin.
4. A clear plastic mask used for oxygen administration that covers the mouth and nose.
8. The windpipe.
11. Children younger than 1 year.
14. The two main branches of the windpipe that lead into the right and left lungs.
16. The two upper chambers of the heart.
18. The lower jaw.
19. The fluid part of the blood that carries blood cells, transports nutrients, and removes cellular waste materials.

Down

1. Heart disease characterized by breathlessness, fluid retention in the lungs, and generalized swelling of the body.
2. The breastbone.
3. The separation of the head from the rest of the body.
5. The passages from the openings of the mouth and nose to the air sacs in the lungs through which air enters and leaves the lungs.
6. The air sacs of the lungs where the exchange of oxygen and carbon dioxide takes place.
7. The smallest blood vessels that connect small arteries and small veins.
8. Burns caused by heat; the most common type of burn.
9. The virus that causes acquired immune deficiency syndrome (AIDS).
10. The process of sorting victims based on the severity of their injuries and their medical needs to establish treatment and transportation priorities.
12. The artificial circulation of the blood and movement of air into and out of the lungs in a pulseless, nonbreathing victim.
13. A surgical opening in the neck that connects the windpipe (trachea) to the skin.
15. Anyone between 1 year of age and the onset of puberty (12 to 14 years of age).
17. An immune disorder caused by infection with the human immunodeficiency virus (HIV), resulting in an increased vulnerability to infections and to certain rare cancers.

Fire Alarms

The following real case scenarios will give you an opportunity to explore the concerns associated with emergency medical care. Read each scenario, and then answer each question in detail.

1. Your company responds to a motor vehicle accident for extrication. Your officer directs you to use universal precautions. What are they?

2. You and your company are rehabilitating after a structure fire. Your partner suddenly slaps his wrist and says he has been stung by a bee. Within minutes, he begins to show signs of labored breathing, complains of an itchy sensation, and has a red tint to his skin.

   A. What may be happening to him?

   B. How should you proceed?

Fire Fighter II in Action

The following scenario will give you an opportunity to apply your firefighting knowledge and your fire department SOGs to the new information you learned while studying this chapter. Research your department's SOGs and answer the assignments in detail. Compare your answers with your classmates' and discuss similarities and obvious differences between your answers.

Your company had been dispatched to an apartment where an elderly gentleman was complaining of difficulty breathing and fever. While treating him, he was continuously coughing, and as you prepared him for transport, he became violently ill. You and your partner were exposed and possibly contaminated by expelled stomach contents as well as airborne droplets from the coughing.

1. What are your department's SOPs on PPE and standard precautions?

2. What are your department's SOPs on decontaminating yourself and limiting any more spread of this type of exposure?

Skill Drills

Skill Drill 25-3: Placing a Victim in the Recovery Position

Test your knowledge of this skill drill by filling in the correct words in the photo captions.

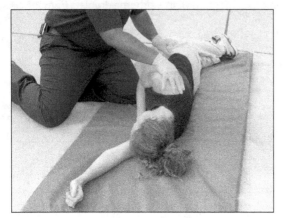

© Jones & Bartlett Learning. Photographed by Glen E. Ellman.

1. Carefully roll the victim onto one side as you support the victim's _____. Roll the victim as a unit without twisting the _____. You can use the victim's _____ to help hold his or her head in a good position.

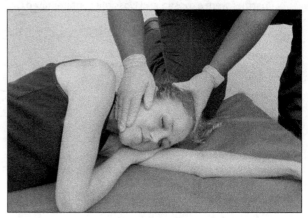

© Jones & Bartlett Learning. Photographed by Glen E. Ellman.

2. Place the victim's head on its side so that any _____ drain out of the _____.

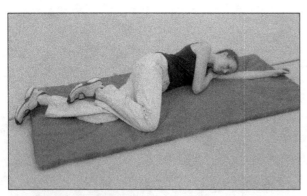

© Jones & Bartlett Learning. Photographed by Glen E. Ellman.

3. Monitor the victim's airway. Bending the victim's _____ will help maintain the victim in the recovery position.

Skill Drill 25-12: Performing One-Rescuer Adult CPR

Test your knowledge of this skill drill by placing the photos below in the correct order. Number the first step with a "1," the second step with a "2," and so on.

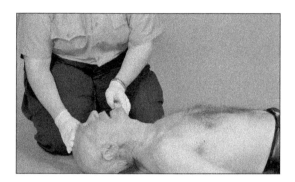

_____ Open the airway.

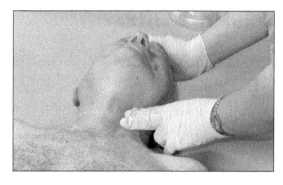

_____ Check for circulation.

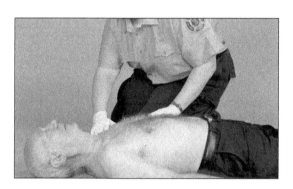

_____ Establish responsiveness and lack of breathing.

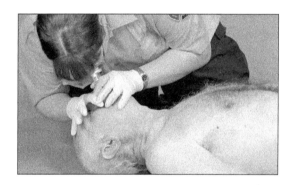

_____ Perform rescue breathing.

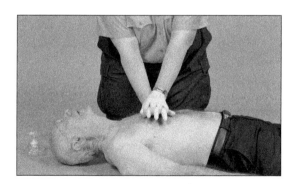

_____ Perform chest compressions.

Vehicle Rescue and Extrication

Workbook Activities

The following activities have been designed to help you. Your instructor may require you to complete some or all of these activities as a regular part of your fire fighter training program. You are encouraged to complete any activity that your instructor does not assign as a way to enhance your learning in the classroom.

Chapter Review

The following exercises provide an opportunity to refresh your knowledge of this chapter.

Matching

Match each of the terms in the left column to the appropriate definition in the right column.

_____ 1. Conventional vehicles
_____ 2. Air chisel
_____ 3. Tempered glass
_____ 4. Bulkhead
_____ 5. Rescue-lift air bags
_____ 6. Step chocks
_____ 7. Cribbing
_____ 8. Wedges
_____ 9. Laminated glass
_____ 10. Posts

A. Specialized cribbing assemblies
B. A tool that is efficient at cutting sheet metal and some plastics
C. Used to snug loose cribbing under the load or when using lift air bags to fill the void between the crib and the object as it is raised
D. The vertical support members of a vehicle that holds up the roof and forms the upright columns of the occupant cage
E. Glass used to make car windshields
F. Glass used for side and rear car windows
G. A vehicle that uses an internal combustion engine
H. Short lengths of usually hardwood timber used to stabilize vehicles
I. Pneumatic-filled bladders made out of rubber or synthetic material
J. The wall that separates the engine compartment from the passenger compartment

Multiple Choice

Read each item carefully, and then select the best response.

_____ 1. Vehicles that are powered by compressed natural gas are known as
   A. electric-powered vehicles.
   B. hybrid vehicles.
   C. alternative-powered vehicles.
   D. conventional vehicles.

_____ 2. Axes, bolt cutters, hacksaws, and manual hydraulic cutters are all tools that can be used for
   A. prying.
   B. stabilizing.
   C. windshield removal.
   D. cutting.

CHAPTER 26

_____ 3. What is the first step in displacing the roof of a vehicle?
   A. Remove the "A" posts.
   B. Cut the "C" posts.
   C. Ensure the safety of the rescuers.
   D. Remove the glass.

_____ 4. As soon as you have secured access to the victim, what is your next step as a rescuer?
   A. Remove the excess glass.
   B. Remove the extraneous materials.
   C. Begin to provide emergency medical care.
   D. Begin to communicate with the victims.

_____ 5. When removing a windshield with an axe, one rescuer begins
   A. at the top, in the middle.
   B. at the driver side.
   C. at the passenger side.
   D. at the bottom.

_____ 6. The victim needs to be stabilized and packaged in preparation for removal in the
   A. stabilization phase.
   B. extrication phase.
   C. size-up phase.
   D. rescue phase.

_____ 7. When removing the roof of a vehicle, it is essential to remove the
   A. windshield.
   B. "C" posts.
   C. doors.
   D. bulkhead.

_____ 8. What is the first step of the dash displacement procedure?
   A. Open both front doors.
   B. Remove the bulkhead.
   C. Remove the steering wheel.
   D. Cut the "A" post.

_____ 9. What is the simplest way to displace a seat backward?
   A. Cut the material out of the bottom of the seat.
   B. Remove the backseat.
   C. Tilt the seat backward.
   D. Move the seat backward in its tracks.

_____ 10. The suspension system of most vehicles can be stabilized with
   A. cribbing.
   B. straps.
   C. wedges.
   D. step chocks.

_____ 11. Between the layers of glass that make the vehicle windshield is
   A. an empty space.
   B. a thin layer of flexible plastic.
   C. an epoxy glue.
   D. an invisible wire mesh.

_____ 12. Cribbing, rescue-lift air bags, and step blocks are all types of
   A. stabilization devices.
   B. patient extractors.
   C. bracing tools.
   D. prying tools.

_____ 13. Which posts are located closest to the front of the vehicle?
   A. "A" posts
   B. "B" posts
   C. "C" posts
   D. "D" posts

_____ 14. The rear window of a vehicle is made of
   A. tempered glass.
   B. laminated windshield glass.
   C. block glass.
   D. sheet glass.

_____ 15. Steering wheels can be cut using
   A. a hacksaw.
   B. a bolt cutter.
   C. a hydraulic cutter.
   D. all of the above.

_____ 16. A vehicle windshield is made of
   A. tempered glass.
   B. laminated windshield glass.
   C. block glass.
   D. sheet glass.

_____ 17. Cribbing protects the vehicle from
   A. electrical hazards.
   B. excessive exposure.
   C. rolling.
   D. other transportation.

_____ 18. What are the most efficient and widely used tools for opening jammed doors?
   A. Cutting tools
   B. Manual hydraulic tools
   C. Powered hydraulic tools
   D. Prying tools

_____ 19. When using rescue-lift air bags, use _____ to fill the void between the crib and the vehicle.
   A. step chocks
   B. wedges
   C. posts
   D. bulkheads

_____ 20. What is the most common type of rescue-lift air bag?
   A. Low-pressure lift air bag
   B. Medium-pressure lift air bag
   C. High-pressure lift air bag
   D. Dual-pressure lift air bag

Labeling

Label the following diagram with the correct terms.

1. Anatomy of a vehicle.

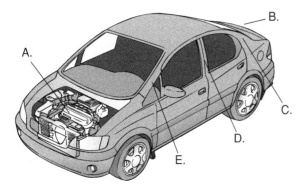

Figure 26-2

A. _____
B. _____
C. _____
D. _____
E. _____

Vocabulary

Define the following terms using the space provided.

1. Hybrid vehicle:

2. Firewall:

3. Platform frame:

4. Post:

5. Unibody:

Fill-in

Read each item carefully, and then complete the statement by filling in the missing word(s).

1. The simplest way to access a victim of a crash is to open a(n) _____.

2. A fire in a natural gas-powered vehicle poses the threat of a(n) _____.

3. The first step in the extrication process is _____.

4. If the door cannot be opened or glass removal will not provide access to the victim, the most common technique for gaining access is _____ displacement.

5. After arriving at the scene of a motor vehicle collision, it is important to assess the _____ present and to determine the _____ of the incident.

6. The right side of the vehicle is where the _____ seat is located.

7. One method of displacing the roof is to cut the _____ posts and fold the roof back toward the rear of the vehicle.

8. Traffic hazards are best handled by the appropriate _____ _____ agency.

9. The three types of commonly used pneumatic rescue-lift air bags are low, medium, and high _____.

10. The _____ posts are located between the front and rear doors of a vehicle.

True/False

If you believe the statement to be more true than false, write the letter "T" in the space provided. If you believe the statement to be more false than true, write the letter "F."

_____ 1. Unibody construction combines the vehicle body and the frame into a single component.

_____ 2. Wedges should be the same width as the cribbing used in stabilization efforts.

_____ 3. Rescue-lift air bags are among the best pieces of equipment used to shore a vehicle by themselves.

_____ 4. The purpose of disentangling the victim is to remove those parts of the vehicle that are trapping the victim.

_____ 5. When it is necessary to force a door to gain access to a victim, choose the door closest to the victim.

_____ 6. The steps of scene stabilization consist of reducing, removing, or mitigating the hazards at the incident scene.

_____ 7. Downed power lines can create a mechanical hazard.

_____ 8. The incident commander will usually perform a size-up of the scene by conducting a 360-degree walk around the scene.

_____ 9. Unstable objects pose a more serious threat to rescuers than do stabilized vehicles.

Short Answer

Complete this section with short written answers using the space provided.

1. Identify and provide an example of each of the four general functions of gaining access and disentangling a victim.

2. Identify five vehicle air bag safety tips.

3. Identify five safety tips for using rescue-lift air bags.

Word Fun

The following crossword puzzle is an activity provided to reinforce correct spelling and understanding of terminology associated with firefighting. Use the clues provided to complete the puzzle. Do not include spaces or punctuation when filling in the puzzle.

Clues

Across

1. One of the vertical support members, or pillars, of a vehicle that holds up the roof and forms the upright columns of the passenger compartment.
2. An electrochemical system that consumes fuel to produce an electric current. The main chemical reaction used in a fuel cell for producing electric power is not combustion, but sources of combustion may be used within the overall fuel cell system such as reformers/fuel processors. (NFPA 70)
5. Vertical support members located between the front and rear doors of a motor vehicle.
6. The frame construction most commonly used in vehicles.
7. The wall that separates the engine compartment from the passenger compartment in a motor vehicle.
10. A vehicle that uses a battery-powered electric motor and an internal combustion engine.
11. Vertical support members that form the sides of the windshield of a motor vehicle.
12. Specialized cribbing assemblies made out of wood or plastic blocks in a step configuration. They are typically used to stabilize vehicles.

Down

1. A type of vehicle frame resembling a ladder, which is made up of two parallel rails joined by a series of cross members. This kind of construction is typically used for luxury vehicles, sport utility vehicles, and all types of trucks.
2. Many fuel cells assembled into a group to produce larger quantities of electricity.
3. A type of safety glass that is heat-treated so that it will break into smaller, less dangerous pieces.
4. Short lengths of timber/composite materials that are used in various configurations to stabilize loads in place or while a load is moving. (NFPA 1006)
8. Material used to tighten or adjust cribbing and shoring systems. (NFPA 1006)
9. Vertical support members located behind the rear doors of a motor vehicle.

Fire Alarms

The following real case scenarios will give you an opportunity to explore the concerns associated with vehicle rescue and extrication. Read each scenario, and then answer each question in detail.

1. You have been dispatched to a vehicle accident on the interstate at 1:15 AM on a rainy night. Your company is first on the scene, and standard operating procedures (SOPs) state your first actions are to protect the scene from traffic and then do a size-up.

    A. How will you protect the scene from approaching traffic?

    B. How will you proceed with your scene size-up?

2. Your company has been dispatched to a motor vehicle accident involving two vehicles with trapped occupants. It is 3:00 PM on a bright, clear day; both vehicles are on their wheels. The incident commander (IC) orders your company to stabilize both vehicles.

    A. Which equipment will you use for vehicle stabilization?

    B. How will you stabilize these vehicles?

Fire Fighter II in Action

The following scenario will give you an opportunity to apply your firefighting knowledge and your fire department SOGs to the new information you learned while studying this chapter. Research your department's SOGs and answer the assignment in detail. Compare your answers with your classmates' and discuss similarities and obvious differences between your answers.

1. Automobile safety features and fuel sources change and improve each year as new vehicle models are developed and released for public sale. As a fire fighter II, how will you keep up with these continuous changes, then apply your knowledge to vehicle rescue and extrication?

Skill Drills

Skill Drill 26-2: Performing a Scene Size-up at a Motor Vehicle Crash

Test your knowledge of this skill drill by placing the photos below in the correct order. Number the first step with a "1," the second step with a "2," and so on.

© Jones & Bartlett Learning. Photographed by Glen E. Ellman.

_____ Position emergency vehicles to protect the crash scene and the rescuers. Take any additional actions needed to prevent further crashes.

© Jones & Bartlett Learning. Photographed by Glen E. Ellman.

_____ Direct personnel to perform initial tasks.

© Jones & Bartlett Learning. Photographed by Glen E. Ellman.

_____ Perform a 360-degree walk-around looking for potential hazards. Look for overhead hazards and hazards under the vehicles, and determine the stabilization needed to prevent further movement of the vehicles involved in the incident.

© Jones & Bartlett Learning. Photographed by Glen E. Ellman.

_____ Determine the number of patients, the severity of their injuries, and the amount of entrapment. Give an updated report and call for additional resources if needed.

Fundamentals of Fire Fighter Skills

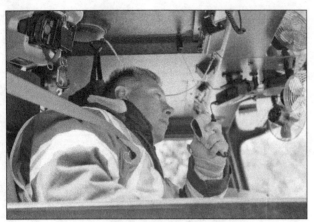

© Jones & Bartlett Learning. Photographed by Glen E. Ellman.

_____ Perform a quick initial assessment from the first-arriving vehicle, establish command, and give a brief initial radio report.

© Jones & Bartlett Learning. Photographed by Glen E. Ellman.

_____ Establish a secure working area and an equipment staging area.

Skill Drill 26-4: Stabilizing a Vehicle Following a Motor Vehicle Crash
Test your knowledge of this skill drill by filling in the correct words in the photo captions.

© Jones & Bartlett Learning. Photographed by Glen E. Ellman.

1. Don personal protective equipment (PPE), including face protection. Minimize hazards to _____ and _____. _____ both sides of one tire to prevent the vehicle from _____ by placing one _____ in front of a wheel and a second _____ in back of the _____.

© Jones & Bartlett Learning. Photographed by Glen E. Ellman.

2. Consider _____ tires for added stability. Assess the need for _____ _____ for additional stability.

© Jones & Bartlett Learning. Photographed by Glen E. Ellman.

3. Turn off the _____ and remove the key or fob.

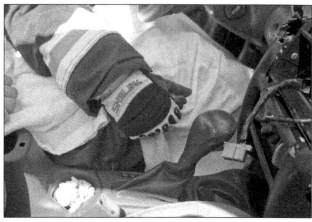
© Jones & Bartlett Learning. Photographed by Glen E. Ellman.

4. Place the gear shift in _____, and apply the _____ _____.

Assisting Special Rescue Teams

Workbook Activities

The following activities have been designed to help you. Your instructor may require you to complete some or all of these activities as a regular part of your fire fighter training program. You are encouraged to complete any activity that your instructor does not assign as a way to enhance your learning in the classroom.

Chapter Review

The following exercises provide an opportunity to refresh your knowledge of this chapter. All questions in this chapter are Fire Fighter II level.

Matching

Match each of the terms in the left column to the appropriate definition in the right column.

_____ 1. High-angle operations

_____ 2. IDLH

_____ 3. Low-angle operations

_____ 4. SAR system

_____ 5. PFD

_____ 6. Shoring

_____ 7. Hot zone

_____ 8. Spoil pile

_____ 9. Warm zone

_____ 10. Placards

A. The area immediately surrounding an incident site that is directly dangerous to life and health

B. Allows the body to float in water

C. The slope of the ground is greater than 45 degrees, and fire fighters are dependent on life safety rope for support

D. Signage, required to be placed on all four sides of vehicles, that identifies the hazardous materials contents being transported by the vehicles

E. Flammable, toxic, or oxygen-deficient atmospheres that pose immediate or possible adverse health effects

F. A method of supporting a trench wall to prevent its collapse

G. An emergency breathing system that utilizes an air line running from the rescuers to a fixed air supply located outside of the confined space

H. When fire fighters are dependent on the ground for their primary support, and the rope system is a secondary means of support

I. The area between the hot zone and the cold zone

J. The unstable pile of dirt removed from an evacuation

Multiple Choice

Read each item carefully, and then select the best response.

_____ 1. To ensure the continuity of quality care and proper transfer of responsibility, there must be
   A. a complete team debriefing.
   B. adequate reporting and accurate records.
   C. an incident analysis report.
   D. a critical incident stress management (CISM) intervention.

CHAPTER 27

_____ 2. A collapse that occurs after an initial collapse is called a
   A. follow-up collapse.
   B. supportive collapse.
   C. shoring collapse.
   D. secondary collapse.

_____ 3. During elevator and escalator rescue, some activities should be attempted only by
   A. technical rescue team members.
   B. emergency responders.
   C. professionally trained service technicians.
   D. veteran department members.

_____ 4. A _____ is an enclosed area that is not designed for people to occupy.
   A. vault
   B. confined space
   C. box car
   D. SAR

_____ 5. In search and rescue, removing a victim from a hostile environment is classified as a
   A. search.
   B. rescue.
   C. recovery.
   D. removal.

_____ 6. Once the victim has been removed from the hazard area, who transports the victim to an appropriate medical facility?
   A. The technical rescue team
   B. The operations team
   C. EMS
   D. The safety officer

_____ 7. By using a personnel _____ and working within the ICS, an IC can track the resources at the scene, make appropriate assignments, and ensure that every person at the scene operates safely.
   A. buddy system
   B. tagout system
   C. accountability system
   D. two-in/two-out rule

_____ 8. What are the most versatile and widely used technical rescue skills?
   A. Disentanglement skills
   B. Medical skills
   C. Rope skills
   D. Hazardous materials knowledge

_____ 9. Warning _____ are required for most hazardous materials in storage or in transit.
   A. lights
   B. sirens
   C. papers
   D. placards.

_____ 10. During a technical rescue incident, whose orders do fire fighters follow?
   A. The company officer's
   B. The incident commander's
   C. The battalion chief's
   D. The rescue captain's

_____ 11. During a rescue incident, emergency medical care should be initiated as soon as
   A. the technical rescue team is clear of the scene.
   B. access is made to the victim.
   C. the medical team arrives on scene.
   D. the incident commander (IC) indicates that medical treatment is required.

_____ 12. The lack of oxygen and the presence of poisonous gases are the greatest hazards associated with a
   A. vehicle or machinery rescue.
   B. high-angle rescue.
   C. hazardous materials rescue.
   D. confined-space rescue.

_____ 13. If a victim's general condition is deteriorating and time will not permit meticulous splinting and dressing procedures, the victim may be removed by
   A. the medical team.
   B. rapid extrication.
   C. complete immobilization.
   D. special extraction teams.

_____ 14. A technical rescue team will usually respond with a rescue squad,
   A. medic unit, and safety officer.
   B. paramedic, and incident commander.
   C. logistics team, and operational team.
   D. medic unit, engine company, and chief.

_____ 15. The area used for staging vehicles and equipment is called the
   A. hot zone.
   B. warm zone.
   C. cold zone.
   D. public zone.

_____ 16. In an industrial setting, securing the scene is the responsibility of the
   A. facility supervisor.
   B. incident commander.
   C. emergency response team.
   D. property owner.

_____ 17. Which level of training allows an individual to work in the warm zone and directly assist those conducting the rescue operation?
   A. Awareness level
   B. Operations level
   C. Technician level
   D. Incident commander level

_____ 18. Shutting off the utilities in the area where the rescuers will be working is a responsibility of the
   A. incident commander.
   B. safety officer.
   C. logistics officer.
   D. shift captain.

_____ 19. The overriding objective for each rescue, transfer, and removal is to complete the process as
   A. quickly as possible.
   B. safely and efficiently as possible.
   C. a team.
   D. an integrated team.

_____ 20. In a vehicle accident, disentanglement is the process of
   A. cutting a vehicle away from the victim.
   B. removing the victim from the vehicle.
   C. cutting and removing the doors of the vehicle.
   D. establishing medical control of the victim.

_____ 21. When responding to an industrial facility, the IC should make contact with the
   A. business owner.
   B. property owner.
   C. responsible party.
   D. city office or administration.

_____ 22. The fire fighter should start compiling the facts about an incident from the
   A. captain.
   B. scene size-up.
   C. initial dispatch of the rescue call.
   D. technical rescue team specialists.

_____ 23. The only time the victim should be moved prior to completion of initial care, assessment, stabilization, and treatment is when there is immediate danger to
   A. the victim's or emergency responder's life.
   B. the surrounding areas.
   C. a rescue team member's life.
   D. the entire rescue operations team.

_____ 24. All emergency service personnel at a rescue situation must
   A. constantly assess and reassess the scene.
   B. communicate with the victim(s).
   C. report directly to the incident commander.
   D. be prepared to assist with the technical rescue team.

_____ 25. What is the most common method of establishing the control zones for an emergency incident site?
   A. Barricades
   B. Pylons
   C. Chalk or paint lines
   D. Fire line tape

Vocabulary

Define the following terms using the space provided.

1. Lockout/tagout system:

2. Hazardous materials:

3. Technical rescue incident:

Fill-in

Read each item carefully, and then complete the statement by filling in the missing word(s).

1. Natural gas and liquefied petroleum gas are nontoxic, but are classified as _____ because they displace breathing air.

2. Information gathered _____ to the technical rescue team's arrival will save valuable time during the actual rescue.

3. _____ collapse is the sudden and unplanned collapse of part or all of a structure.

4. Once the rescue is complete, the scene must be _____ by the rescue crew to ensure that no one else becomes injured.

5. If you have the role of assisting a technical rescue team, _____ with the team is probably the most important thing you can do.

6. Scene control activities are sometimes assigned to _____ _____ personnel.

7. To ensure the safety of the rescuers there must be a(n) _____ _____ _____ in place.

8. The _____ level of training provides an emphasis on recognizing the hazards, securing the scene, and calling for appropriate assistance.

9. It is extremely important that hazardous materials incident victims are _____ prior to transport.

10. To assist in more efficient communication with other rescuers, it is important to know the _____ used in the field.

11. The process of preparing the victim for transport is called _____.

12. A rescue area is an area that surrounds the incident site and whose size is _____ to the hazards that exist.

True/False

If you believe the statement to be more true than false, write the letter "T" in the space provided. If you believe the statement to be more false than true, write the letter "F."

_____ 1. Without a solid command structure, most large-scale rescue efforts are doomed to failure.

_____ 2. Rescue efforts often require a small, focused group of individuals to complete the operation.

_____ 3. During water rescue incidents, all responders within 10 feet (3 meters) of the water must wear an approved personal flotation device.

_____ 4. During a rescue, a team member should remain with the victim to direct the rescuers performing disentanglement.

_____ 5. Any machine that is involved in a machinery rescue should be considered electrically charged.

_____ 6. The most important part of any rescue is the arrival and size-up of the incident scene.

_____ 7. To assist a victim in remaining calm, you should communicate calmly, at a level that the victim can understand.

_____ 8. Tagout procedures are used for personnel accountability.

_____ 9. The best way to prepare for the next rescue call is to review the last one.

_____ 10. Rescue situations have many hidden hazards.

_____ 11. Many fire departments are run like a military organization.

Short Answer

Complete this section with short written answers using the space provided.

1. List five considerations to keep in mind during size-up.

2. List the five guidelines that a fire fighter should follow when assisting rescue team members.

3. Identify the paramilitary guidelines for which a fire fighter must have a strong appreciation in order to understand the command and control concept of fire departments.

4. Identify the 10 steps of special rescue.

5. Identify the components of the acronym "FAILURE" used to describe why rescuers fail.

6. Identify the types of special rescues encountered by fire fighters.

7. Describe how to approach and assist at a confined space rescue incident safely.

8. Describe how to safely approach and assist at a hazardous materials rescue incident.

Word Fun

The following crossword puzzle is an activity provided to reinforce correct spelling and understanding of terminology associated with firefighting. Use the clues provided to complete the puzzle. Do not include spaces or punctuation when filling in the puzzle.

CLUES

Across

6. The process of securing a victim in a transfer device, with regard to existing and potential injuries or illness, so as to prevent further harm during movement. (NFPA 1006)
7. A complex rescue incident requiring specially trained personnel and special equipment to complete the mission. (NFPA 1670)
8. The area located between the hot zone and the cold zone at an incident. The decontamination corridor is located in this zone.
9. A pile of excavated soil next to the excavation or trench. (NFPA 1006)
11. The control zone of an incident that contains the command post and such other support functions as are deemed necessary to control the incident. (NFPA 1500)

Down

1. A structure such as a metal hydraulic, pneumatic/mechanical, or timber system that supports the sides of an excavation and is designed to prevent cave-ins. (NFPA 1670)
2. An area large enough and so configured that a member can bodily enter and perform assigned work, but which has limited or restricted means for entry and exit and is not designed for continuous human occupancy. (NFPA 1500)
3. Signage required to be placed on all four sides of highway transport vehicles, railroad tank cars, and other forms of hazardous materials transportation that identifies the hazardous contents of the vehicle, using a standardized system: 10 inch by 10 inch (25.4 centimeter by 25.4 centimeter) diamond-shaped indicators.
4. The area immediately surrounding a hazardous materials spill/incident site that is directly dangerous to life and health. All personnel working in this zone must wear complete, appropriate protective clothing and equipment. Entry requires approval by the incident commander or a designated sector officer. Complete backup, rescue, and decontamination teams must be in place at the perimeter before operations begin.
5. A condition in which a victim is trapped by debris, soil, or other material and is unable to extricate himself or herself.
10. A displacement device worn to keep the wearer afloat in water. (NFPA 1925)

Fire Alarms

The following real case scenarios will give you an opportunity to explore the concerns associated with assisting special rescue teams. Read each scenario, and then answer each question in detail.

1. Your ladder has been dispatched to an incident involving an individual trapped in industrial machinery. Your company officer has ordered you to assemble the tools that may be needed to release the victim from the machinery while he performs a scene size-up. Which tools will you assemble for use?

   _____
   _____
   _____
   _____
   _____
   _____

2. Your company is dispatched to a trench collapse where a worker is trapped. How will you make a safe approach to the collapse zone?

   _____
   _____
   _____
   _____
   _____
   _____

3. Your company has been dispatched to a call of wires down, with an individual unconscious and in contact with the downed wires. There are several citizens in the immediate area. Your company officer has ordered you to secure the scene and establish a barrier. How will you accomplish these orders?

   _____
   _____
   _____
   _____
   _____
   _____

Fire Fighter II in Action

The following scenarios will give you an opportunity to apply your firefighting knowledge and your fire department SOGs to the new information you learned while studying this chapter. Research your department's SOGs and answer the assignments in detail. Compare your answers with your classmates' and discuss similarities and obvious differences between your answers.

1. Create a list of rescue tools and hand tools you could be carrying in your turnout gear. Keep in mind usefulness as well as weight and bulk.

2. What additional training would be important to you as a fire fighter II if you lived near a river that was prone to flooding?

Hazardous Materials: Overview

Workbook Activities

The following activities have been designed to help you. Your instructor may require you to complete some or all of these activities as a regular part of your fire fighter training program. You are encouraged to complete any activity that your instructor does not assign as a way to enhance your learning in the classroom.

Chapter Review

The following exercises provide an opportunity to refresh your knowledge of this chapter.

Matching

Match each of the terms in the left column to the appropriate definition in the right column.

_____ 1. NFPA
_____ 2. Hazardous material
_____ 3. MSDS
_____ 4. Awareness level
_____ 5. Target hazard
_____ 6. SERC
_____ 7. Technician level
_____ 8. Hazardous waste
_____ 9. SARA
_____ 10. EPA

A. Law that affects how fire departments respond in a hazardous materials emergency
B. Training that provides the ability to enter heavily contaminated areas using the highest levels of protection
C. A detailed profile of a single chemical or mixture of chemicals provided by the manufacturer or supplier of a chemical
D. A federal agency that regulates and governs issues relating to hazardous materials in the environment
E. Any person who comes upon an incident and has been trained to recognize, identify, and notify
F. The body that develops and maintains nationally recognized minimum consensus standards on many areas of fire safety and hazardous materials
G. The impure substance left after manufacturing
H. A facility that presents a high potential for loss of life
I. Liaison between local and state levels of emergency response authorities
J. Any material that poses an unreasonable risk of damage or injury to persons, property, or the environment if not properly controlled

Multiple Choice

Read each item carefully, and then select the best response.

_____ 1. In the United States, which federal government agency enforces and publicizes laws and regulations that govern the transportation of goods by highways, rail, and air?
   A. Environmental Protection Agency (EPA)
   B. Occupational Safety and Health Agency (OSHA)
   C. State Emergency Response Commission (SERC)
   D. Department of Transportation (DOT)

CHAPTER 28

_____ 2. Which response level is trained to take offensive actions?
   A. Awareness
   B. Operations
   C. Technician
   D. Specialist

_____ 3. A material that poses an unreasonable risk to the health and safety of the public and/or the environment if it is not controlled properly during handling, processing, and disposal is called a
   A. hazardous waste.
   B. hazardous material.
   C. hazardous target.
   D. hazardous substance.

_____ 4. Which of the following is a state group that acts as a liaison between local- and state-level response authorities?
   A. SARA
   B. EPA
   C. SERC
   D. MSDS

_____ 5. Which response level is trained to take defensive actions?
   A. Awareness
   B. Operations
   C. Technician
   D. Specialist

_____ 6. Which of the following is a nongovernment agency that issues fire response standards?
   A. CANUTEC
   B. NFPA
   C. OSHA
   D. EPA

_____ 7. Which response level is trained to recognize a hazardous materials emergency and call for assistance?
   A. Awareness
   B. Operations
   C. Technician
   D. Specialist

_____ 8. In the United States, the federal document containing the hazardous materials response competencies is known as
   A. NFPA.
   B. EPCRA.
   C. SARA.
   D. HAZWOPER.

Fundamentals of Fire Fighter Skills

_____ **9.** Which of the following is a group that gathers information about hazardous materials and disseminates that information to the public?
   A. LEPC
   B. NFPA
   C. SARA
   D. EPA

_____ **10.** What act requires a business that handles chemicals to report storage type, quantity, and storage methods to the fire department and the local emergency planning committee?
   A. Superfund Amendments and Reauthorization Act
   B. Local Emergency Planning Committee Act
   C. Emergency Planning and Community Right to Know Act
   D. Occupational Safety and Health Act

Vocabulary

Define the following terms using the space provided.

1. Specialist level:

2. Local emergency planning committee (LEPC):

3. Operations level:

4. Material safety data sheet (MSDS):

5. HAZWOPER:

Fill-in

Read each item carefully, and then complete the statement by filling in the missing word(s).

1. _____ activities enable agencies to develop logical and appropriate response procedures for anticipated incidents.

2. _____ _____ is the material that remains after a manufacturing plant has used some chemicals, and they are no longer pure.

3. The _____ _____ _____ regulates and governs issues relating to hazardous materials in the environment.

4. States that have adopted OSHA safety and health regulations are called _____-_____ states.

5. Hazardous materials incidents are _____ complicated than most structural firefighting incidents.

6. The federal document containing the hazardous materials response competencies is known as _____.

7. Awareness level skills are _____, _____, and _____.

8. _____ _____ is the *Standard for Competence of Responders to Hazardous Materials/Weapons of Mass Destruction Incidents*.

9. A(n) _____ _____ _____ _____ is a detailed profile of a single chemical or mixture of chemicals, provided by the manufacturer and/or supplier of a chemical.

10. Hazardous materials _____ receive the most aggressive level of training as identified in NFPA 472.

True/False

If you believe the statement to be more true than false, write the letter "T" in the space provided. If you believe the statement to be more false than true, write the letter "F."

_____ 1. The SARA regulates and governs issues relating to hazardous materials and the environment.

_____ 2. The actions taken at hazardous materials incidents are largely dictated by the chemicals involved in the incident.

_____ 3. The ability to recognize a potential hazardous materials incident is critical to ensuring one's safety.

_____ 4. The EPA's version of HAZWOPER is in Title 40, Protection of the Environment, Part 311, Worker Safety.

_____ 5. Each state has a State Emergency Response Commission (SERC) that acts as a liaison between local and state levels of authority.

_____ 6. The goal of a fire fighter is to favorably change the outcome of a hazardous materials incident.

_____ 7. Fires require a less straightforward response than do hazardous materials incidents.

_____ 8. Response agencies should not preplan target hazards owing to the health issues involved in such planning.

_____ 9. The Emergency Planning and Community Right to Know Act was one of the first laws to affect how fire departments respond in a hazardous materials emergency.

_____ 10. When approaching a hazardous materials event, you should make a conscious effort to change your response perspective.

Short Answer

Complete this section with short written answers using the space provided.

1. Discuss the Superfund Amendments and Reauthorization Act.

   _____
   _____
   _____
   _____
   _____
   _____
   _____

2. Identify and describe the five levels of hazardous materials training and competencies, according to NFPA 472.

   _____
   _____
   _____
   _____
   _____
   _____
   _____

Word Fun

The following crossword puzzle is an activity provided to reinforce correct spelling and understanding of terminology associated with firefighting. Use the clues provided to complete the puzzle. Do not include spaces or punctuation when filling in the puzzle.

CLUES

Across

5. The liaison between local and state levels that collects and disseminates information relating to hazardous materials. This involves agencies such as the fire service, police services, and elected officials.
6. The OSHA regulation that governs hazardous materials waste sites and response training. Specifics can be found in 29 CFR 1910.120. Subsection (q) is specific to emergency response.
8. Established in 1970, the federal agency that ensures safe manufacturing, use, transportation, and disposal of hazardous substances.
9. One of the first federal laws to affect how fire departments respond in a hazardous material emergency.
10. The federal agency that regulates worker safety and, in some cases, responder safety. It is a part of the Department of Labor.
11. Persons who respond to hazardous materials/weapons of mass destruction incidents for the purpose of implementing or supporting actions to protect nearby persons, the environment, or property from the effects of the release. (NFPA 472)
13. The federal agency that publicizes and enforces rules and regulations that relate to the transportation of many hazardous materials.
14. Persons who respond to hazardous materials/weapons of mass destruction incidents who have received more specialized training than hazardous materials technicians. Most of the training that these employees receive is either product or transportation mode specific.

Down

1. Any occupancy type or facility that presents a high potential for loss of life or serious impact to the community resulting from fire, explosion, or chemical release.
2. Personnel who, in the course of their normal duties, could encounter an emergency involving hazardous materials and weapons of mass destruction (WMDs) and who are expected to recognize the presence of the hazardous materials and WMDs, protect themselves, call for trained personnel, and secure the scene. (NFPA 472)
3. Waste that is potentially damaging to the environment or human health due to its toxicity, ignitability, corrosivity, chemical reactivity, or another cause. (NFPA 820)
4. A group comprising members of industry, transportation, the public at large, media, and fire and police agencies; it gathers and disseminates information on hazardous materials stored in the community and ensures that there are adequate local resources to respond to a chemical event in the community.
7. Transitioning to Safety Data Sheets (SDS). A form, provided by manufacturers and compounders (blenders) of chemicals, containing information about chemical composition, physical and chemical properties, health and safety hazards, emergency response, and waste disposal of the material. (NFPA 472)
12. A private organization that develops and maintains nationally recognized minimum consensus standards on many areas of fire safety and specific standards on hazardous materials.

Fire Alarms

The following real case scenarios will give you an opportunity to explore the concerns associated with hazardous materials. Read each scenario, and then answer each question in detail.

1. You have just completed your hazardous materials training. Your Lieutenant gives you an assignment to complete a mock preincident plan for a hazardous materials response to an occupancy near your station. How should you proceed?

   _____
   _____
   _____
   _____
   _____
   _____

2. You are participating in a hazardous materials awareness class, and the instructor asks you to define the term "hazardous materials." What is your response?

   _____
   _____
   _____
   _____
   _____
   _____

Fire Fighter II in Action

The following scenario will give you an opportunity to apply your firefighting knowledge and your fire department SOGs to the new information you learned while studying this chapter. Research your department's SOGs and answer the assignment in detail. Compare your answers with your classmates' and discuss similarities and obvious differences between your answers.

You have just been informed that your rural metro bus line is planning to change fuels. They will be switching from diesel engines to natural gas burning engines.

1. What are your response concerns to accidents involving these new buses?

   _____
   _____
   _____
   _____
   _____
   _____

2. How will you gather more information about these buses and the new refueling station that will be built to accommodate the new fuel?

Hazardous Materials: Properties and Effects

Workbook Activities

The following activities have been designed to help you. Your instructor may require you to complete some or all of these activities as a regular part of your fire fighter training program. You are encouraged to complete any activity that your instructor does not assign as a way to enhance your learning in the classroom.

Chapter Review

The following exercises provide an opportunity to refresh your knowledge of this chapter.

Matching
Match each of the terms in the left column to the appropriate definition in the right column.

_____ 1. Expansion ratio
_____ 2. Toxicology
_____ 3. Carcinogen
_____ 4. Vapor
_____ 5. Vapor pressure
_____ 6. Corrosivity
_____ 7. Flammable range
_____ 8. Pulmonary edema
_____ 9. Lower flammable limit
_____ 10. Vapor density

A. A cancer-causing agent
B. The weight of a gas as compared to an equal volume of dry air
C. A description of a volume increase that occurs when a liquid changes to a gas
D. The gas phase of a substance
E. The study of the adverse effects of chemical or physical agents on living organisms
F. The minimum amount of gaseous fuel that must be present in the air mixture for the mixture to be flammable or explosive
G. Fluid buildup in the lungs
H. The pressure exerted by a liquid's vapor until the liquid and vapor reach an equilibrium
I. The boundaries of a fuel/air mixture necessary for a combustible material to burn properly
J. The ability of a material to cause damage upon skin contact

Multiple Choice
Read each item carefully, and then select the best response.

_____ 1. Bases have pH values that are
   A. equal to zero.
   B. greater than 7.
   C. equal to 7.
   D. less than 7.

_____ 2. Which is the least penetrating of the three types of radiation?
   A. Alpha particles
   B. Beta particles
   C. Gamma radiation
   D. Neutrons

_____ 3. Which type of chemical causes a substantial proportion of exposed people to develop an allergic reaction in normal tissue after repeated exposure to that chemical?
   A. Sensitizer
   B. Irritant
   C. Convulsant
   D. Contaminant

_____ 4. Exposure to which of the following substances prevents the body from using oxygen?
   A. Chlorine
   B. Cyanide
   C. Lewisite
   D. Sarin

_____ 5. Adverse health effects caused by long-term exposure to a substance are termed
   A. acute health hazards.
   B. chronic health hazards.
   C. long-term disablers.
   D. overexposure.

_____ 6. Phosgene is a
   A. nerve agent.
   B. blistering agent.
   C. choking agent.
   D. blood agent.

_____ 7. pH is an expression of the concentration of
   A. hydrogen ions in a given substance.
   B. acid ions in a given substance.
   C. oxygen ions in a given substance.
   D. base ions in a given substance.

_____ 8. The hazardous chemical compounds released when a material decomposes under heat are known as
   A. carcinogens.
   B. alpha particles.
   C. toxic products of combustion.
   D. beta particles.

_____ 9. The characteristics of a chemical that are measurable are
   A. physical properties.
   B. chemical properties.
   C. neither chemical nor physical properties.
   D. both chemical and physical properties.

_____ 10. The temperature at which a liquid changes into a gas is the
   A. flash point.
   B. vaporization point.
   C. boiling point.
   D. gas point.

_____ 11. Which type of exposure occurs when harmful substances are brought into the body through the respiratory system?
   A. Ingestion exposure
   B. Inhalation exposure
   C. Absorption exposure
   D. Injection exposure

_____ 12. The nucleus of a radioactive isotope includes an unstable configuration of
   A. protons and neutrons.
   B. electrons and protons.
   C. electrons and neutrons.
   D. protons, electrons, and neutrons.

_____ 13. The process by which a person or object transfers contamination to another person or object by direct contact is called
   A. contamination by association.
   B. secondary exposure.
   C. direct contamination.
   D. secondary contamination.

_____ 14. The expansion ratio is a description of the volume increase that occurs when a material changes from
   A. a liquid to a solid.
   B. a solid to a gas.
   C. a solid to a liquid.
   D. a liquid to a gas.

_____ 15. The vapor pressure at the standard atmospheric pressure of 20°C can be expressed in pounds per square inch, atmospheres, and millimeters of mercury as follows:
   A. 14.7 psi = 1 atm = 760 mm Hg.
   B. 1 psi = 0.59 atm = 760 torr = 10 mm Hg.
   C. 10 psi = 1 atm = 550 torr = 0.59 mm Hg.
   D. 14.7 psi = 100 atm = 30 torr = 100 mm Hg.

_____ 16. Common acids have pH values that are
   A. equal to zero.
   B. greater than 7.
   C. equal to 7.
   D. less than 7.

_____ 17. The ability of a substance to dissolve in water is known as its
   A. expansion ratio.
   B. dissolvability.
   C. water solubility.
   D. dispersement value.

_____ 18. The first step in understanding the hazard of any chemical involves identifying
   A. physical properties.
   B. chemical properties.
   C. states of matter.
   D. radiation agents.

_____ 19. The ability of a chemical to undergo a change in its chemical makeup, usually with a release of some form of energy, is a
   A. property change.
   B. physical change.
   C. chemical change.
   D. change of state.

_____ 20. Air has a set vapor density value of
   A. 0.59.
   B. 1.0.
   C. 2.4.
   D. 3.8.

___ **21.** The weight of an airborne concentration as compared to an equal volume of dry air is the
   A. vapor density.
   B. vapor ratio.
   C. flammable range.
   D. explosive ratio.

___ **22.** Steel rusting and wood burning are examples of
   A. physical changes.
   B. chemical changes.
   C. vaporization.
   D. ionization.

Labeling

Label the following diagrams with the correct terms.

1. Vapor density.

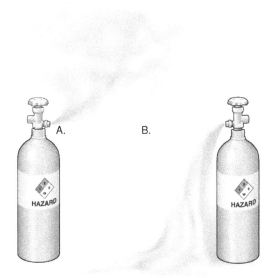

A. _____

B. _____

Figure 29-3

2. Alpha, beta, and gamma radiation.

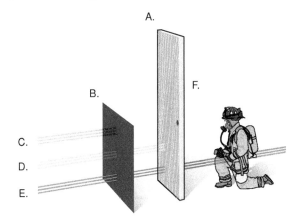

A. _____

B. _____

C. _____

D. _____

E. _____

F. _____

Figure 29-6

3. The four ways a chemical substance can enter the body.

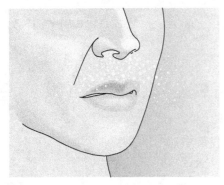
A.

B.

C.

D.

Figure 29-8

A. _____
B. _____
C. _____
D. _____

Vocabulary

Define the following terms using the space provided.

1. Contamination:

2. Flammable vapor:

3. Weapon of mass destruction (WMD):

4. Radiation:

5. HEPA filter:

Fundamentals of Fire Fighter Skills

Fill-in
Read each item carefully, and then complete the statement by filling in the missing word(s).

1. The _____ temperature is the minimum temperature at which a substance will ignite without an external ignition source.

2. If the state of matter and physical properties of a chemical are known, a fire fighter can _____ what the substance will do if it escapes its container.

3. _____ _____ is the minimum temperature at which a liquid or a solid emits vapor sufficient to form an ignitable mixture with air.

4. Chemicals that are capable of causing seizures are classified as _____.

5. Chemicals can undergo a(n) _____ change when subjected to outside influences such as heat, cold, and pressure.

6. _____ particles can break chemical bonds, creating ions; therefore they are considered ionizing radiation.

7. The flash point of gasoline is _____.

8. Standard atmospheric pressure at sea level is _____ pounds per square inch.

9. Most flammable liquids will _____ on water.

10. The periodic table illustrates all the known _____ that make up every known compound.

True/False

If you believe the statement to be more true than false, write the letter "T" in the space provided. If you believe the statement to be more false than true, write the letter "F."

_____ 1. Vapor pressure directly correlates to the speed at which a material will evaporate once it is released from its container.

_____ 2. A chemical brought into the body through an open cut is an injection exposure.

_____ 3. The nucleus of an atom is made up of protons, neutrons, and electrons.

_____ 4. A physical change is essentially a change in state; a chemical change results in an alteration of the chemical nature of the material.

_____ 5. Diesel fuel has a higher flash point than does gasoline.

_____ 6. Nerve agents attack the central nervous system.

_____ 7. Radioactive isotopes can be detected by the noise and odors they give off.

_____ 8. The wider the flammable range, the more dangerous the material.

_____ 9. A hazard is a material capable of posing an unreasonable risk to health, safety, or the environment.

_____ 10. Water has an expansion rate of 100:1 and a boiling point of 100°F.

Short Answer

Complete this section with short written answers using the space provided.

1. What does "HA HA MICEN" stand for?

2. The health hazards posed by radiation are a function of what two factors?

3. Identify the nerve agent signs and symptoms represented by the mnemonic "SLUDGEM."

4. Identify and define the four ways through which chemical substances can enter the human body.

5. Identify five types of possible hazardous materials incidents represented by the mnemonic "TRACEMP."

Word Fun

The following crossword puzzle is an activity provided to reinforce correct spelling and understanding of terminology associated with firefighting. Use the clues provided to complete the puzzle. Do not include spaces or punctuation when filling in the puzzle.

CLUES

Across

2 The study of the adverse effects of chemical or physical agents on living organisms.
6 The highest concentration of a combustible substance in a gaseous oxidizer that will propagate a flame.
7 A fuel complex defined by kind, arrangement, volume, condition, and location that determines the ease of ignition and/or of resistance to fire control.
11 A nerve agent that when dispersed sends droplets in the air that when inhaled harm intended victims.
13 A cancer-causing substance.
14 The minimum temperature at which a liquid emits vapor in sufficient concentration to form an ignitable mixture with air near the surface of the liquid.
15 The lowest temperature at which a liquid will ignite and achieve sustained burning when exposed to a test flame.
16 The gas phase of a substance, particularly of those that are normally liquids or solids at ordinary temperatures.

Down

1 The process by which people, animals, the environment, and equipment are subjected to or come into contact with a hazardous material.
3 The minimum concentration of combustible vapor or combustible gas in a mixture of the vapor or gas and gaseous oxidant above which propagation of flame will occur on contact with an ignition source.
4 The spontaneous decay or disintegration of an unstable atomic nucleus accompanied by the emission of radiation.
5 A chemical that causes a large portion of people or animals to develop an allergic reaction after repeated exposure to the substance.
8 A material with a pH value less than 7.
9 Exposure to a hazardous material by breathing it into the lungs.
10 A yellowish gas that is about 2.5 times heavier than air and slightly water-soluble. It has many industrial uses but also damages the lungs when inhaled; it is a choking agent.
12 A penetrating particle found in the nucleus of the atom that is removed through nuclear fusion or fission. Although these particles are not radioactive, exposure to them can create radiation.

Fire Alarms

The following real case scenarios will give you an opportunity to explore the concerns associated with hazardous materials. Read each scenario, and then answer each question in detail.

1. It is 2:00 in the afternoon on a Saturday when your engine is dispatched to a shopping mall for a hazardous materials incident. Approximately 50 people have been exposed to a substance that is causing pain and burning to their skin and eyes. Last week, the U.S. terror alert system was activated to level red and a warning was posted for possible attacks on shopping malls and other highly populated buildings. How should you proceed?

   _____
   _____
   _____
   _____
   _____
   _____

2. You are dispatched to the local hospital for a hazardous materials incident. When your engine arrives at the scene, a lab technician states that she believes that one of the medical containers is leaking radioactive material. How should you proceed?

   _____
   _____
   _____
   _____
   _____
   _____

Chapter 29: Hazardous Materials: Properties and Effects

Fire Fighter II in Action

The following scenario will give you an opportunity to apply your firefighting knowledge and your fire department SOGs to the new information you learned while studying this chapter. Research your department's SOGs and answer the assignment in detail. Compare your answers with your classmates' and discuss similarities and obvious differences between your answers.

You have been dispatched to a local "big box" store to investigate a possible spill in the pesticide/herbicide aisle. You and your crew have just completed HazMat Ops level training. As you pull up to the front doors, you notice several people sitting and lying on the sidewalk beside the exit. All are coughing, nauseous, and complaining of eye, nose, and lung irritation.

1. What will your initial scene size-up be?

2. What assistance will you immediately request from your dispatch center?

3. What will your first actions be to attempt to remedy the incident?

Hazardous Materials: Recognizing and Identifying the Hazards

Workbook Activities

The following activities have been designed to help you. Your instructor may require you to complete some or all of these activities as a regular part of your fire fighter training program. You are encouraged to complete any activity that your instructor does not assign as a way to enhance your learning in the classroom.

Chapter Review

The following exercises provide an opportunity to refresh your knowledge of this chapter.

Matching
Match each of the terms in the left column to the appropriate definition in the right column.

_____ 1. Vent pipes   A. Information on a pesticide label that indicates the relative toxicity of the material
_____ 2. Signal words   B. Portable tanks characterized by a unique style of construction
_____ 3. Bill of lading C. Containers designed to preserve the temperature of the cold liquid held inside
_____ 4. Totes   D. One or more small openings in closed-head drums
_____ 5. Drums   E. Shipping papers for roads and highways
_____ 6. Waybill   F. Inverted J-shaped tubes that allow for pressure relief from the pipeline
_____ 7. Dewar containers G. Barrel-like containers
_____ 8. Bungs   H. Shipping papers for trains
_____ 9. Cylinder   I. A portable compressed-gas container
_____ 10. Consist   J. The list of every car on a train

Multiple Choice
Read each item carefully, and then select the best response.

_____ 1. Within the NFPA hazard identification system, which number is used to identify materials that can cause death after a short exposure?
   A. 4
   B. 2
   C. 1
   D. 0

_____ 2. What is the type of clandestine lab most commonly encountered by fire fighters?
   A. Paint lab
   B. Drug lab
   C. Chemical lab
   D. Biological lab

CHAPTER 30

_____ 3. Solids and powders are often stored in
   A. drums.
   B. boxes.
   C. bags.
   D. carboys.

_____ 4. Liquid bulk storage containers have an internal capacity of more than
   A. 500 gallons (1892 liters).
   B. 250 gallons (946 liters).
   C. 182 gallons (689 liters).
   D. 119 gallons (450 liters).

_____ 5. Which type of packaging has an inner containment vessel of glass, plastic, or metal and packaging materials made of rubber or vermiculite?
   A. Type A
   B. Type B
   C. Type C
   D. Type D

_____ 6. What are the 10¾-inch diamond-shaped indicators that must be placed on all four sides of hazardous materials transportation called?
   A. MSDS markers
   B. Labels
   C. Placards
   D. AERG tags

_____ 7. If a radiation incident is expected at a fixed facility, which person should be contacted for information?
   A. The safety officer
   B. The incident commander
   C. The radiation safety officer
   D. The shift supervisor

_____ 8. Shipping papers on a marine vessel are referred to as
   A. waybills.
   B. dangerous cargo manifests.
   C. consists.
   D. freight bills.

_____ 9. Which high-pressure vessels have internal pressures of several hundred pounds per square inch and carry liquefied propane?
   A. IM-101s
   B. IM-102s
   C. IMO-Type 10 containers
   D. IMO-Type 5 containers

_____ 10. Which DOT packaging group designation is used to represent the highest level of danger?
   A. Packaging group I
   B. Packaging group II
   C. Packaging group III
   D. Packaging group V

_____ 11. When large-volume horizontal tanks are stored above ground, they are referred to as
   A. OSTs.
   B. ASTs.
   C. USTs.
   D. GSTs.

_____ 12. One of the most common chemical tankers is a gasoline tanker, also known as a(n)
   A. MC-331 pressure cargo tanker.
   B. MC-307 chemical hauler.
   C. MC-306 flammable liquid tanker.
   D. tube trailer.

_____ 13. Drum bungs can be removed using a
   A. bung wrench.
   B. drum ratchet.
   C. cinching wrench.
   D. drum ring.

_____ 14. Which types of containers are generally V-shaped with rounded sides and are used to carry grain or fertilizers?
   A. Consist tankers
   B. Dry bulk cargo tankers
   C. Carboys
   D. ASTs

_____ 15. Solid bulk storage containers have an internal capacity of more than
   A. 1118 pounds (507 kg).
   B. 1084 pounds (492 kg).
   C. 919 pounds (417 kg).
   D. 882 pounds (400 kg).

_____ 16. Gaseous substances that have been chilled until they liquefy are classified as
   A. cryogenic gases.
   B. crystals.
   C. Dewar liquids.
   D. cryogenic liquids.

_____ 17. Fire fighters should be able to recognize the three basic railcar configurations of
   A. nonpressurized, pressurized, and special use.
   B. dry bulk, liquid, and hazardous materials.
   C. agricultural, mechanical, and products.
   D. contained, unpackaged, and hazardous materials.

_____ 18. Within the NFPA hazard identification system, which number is used to identify materials that will not burn?
   A. 4
   B. 2
   C. 1
   D. 0

_____ 19. Propane cylinders contain a liquefied gas and have low pressures of approximately
   A. 50–110 psi (345–758 kPa).
   B. 150–200 psi (1034–1379 kPa).
   C. 200–300 psi (1379–2068 kPa).
   D. 300–400 psi (2068–4137 kPa).

_____ 20. A glass, plastic, or steel container that holds 5 to 15 gallons (19 to 57 liters) of product is a
   A. carboy.
   B. bottle.
   C. drum.
   D. cylinder.

Labeling

Label the following diagram with the correct terms.

1. Chemical transport vehicles.

A. _____

B. _____

C. _____

D. _____

E. _____

F. _____

G. _____

H. _____

Figure 30-6

Vocabulary

Define the following terms using the space provided.

1. Pipeline right-of-way:

   _____
   _____
   _____

2. Shipping papers:

   _____
   _____
   _____

3. Placards and labels:

   _____
   _____
   _____

4. Hazardous materials:

   _____
   _____
   _____

5. Secondary containment:

   _____
   _____
   _____

Fill-in

Read each item carefully, and then complete the statement by filling in the missing word(s).

1. Within the NFPA hazard identification system, special hazards that react with water are identified by _____.

2. Shipping papers for railroad transportation are called _____; the list of every car on the train is called a(n) _____.

3. The _____ _____ _____ _____ describes the chemical hazards posed by a particular substance and provides guidance about personal protective equipment employees need to use to protect themselves from workplace hazards.

4. The _____ marking system has been developed primarily to identify detonation, fire, and special hazards.

5. MC-_____ corrosives tankers are used for transporting concentrated nitric acids and other corrosive substances.

6. A common source of information about a particular chemical is the _____ _____ _____ _____ specific to that substance.

7. Scene _____-_____ is especially important in all hazardous materials incidents.

8. Large-diameter _____ transport natural gas, diesel fuel, and other products from delivery terminals to distribution facilities.

9. IM-_____ containers primarily carry flammable liquids and corrosives.

10. Compressed gases such as hydrogen, oxygen, and methane are carried by _____ trailers.

True/False

If you believe the statement to be more true than false, write the letter "T" in the space provided. If you believe the statement to be more false than true, write the letter "F."

_____ 1. CHEMRESPECT is a free service that connects fire fighters with chemical manufacturers, chemists, and other product specialists who can help during a chemical incident.

_____ 2. The Department of Transportation's marking system is characterized by a system of signs, colors, and numbers.

_____ 3. More than 4 billion tons of hazardous materials are shipped annually in the United States.

_____ 4. Intermodal tanks are both shipping and storage vehicles.

_____ 5. An MSDS will usually include a responsible-party contact.

_____ 6. Hazardous materials incidents can occur almost anywhere.

_____ 7. Nonbulk storage vessels can also be used as intermodal tanks.

_____ 8. Hazardous materials can be transported in cardboard drums or paper bags.

_____ 9. *ACID* is used to represent acid in the NFPA hazard identification system.

_____ 10. *OX* is used to represent compressed oxygen in the NFPA hazard identification system.

Short Answer

Complete this section with short written answers using the space provided.

1. Identify the nine ERG chemical families.

2. List five pieces of information normally included on a material safety data sheet.

3. Describe the parts and purpose of the NFPA 704 hazard identification system.

4. Identify and describe the four colored sections of the ERG.

5. List five pieces of specific information included on a pesticide bag label.

Word Fun

The following crossword puzzle is an activity provided to reinforce correct spelling and understanding of terminology associated with firefighting. Use the clues provided to complete the puzzle. Do not include spaces or punctuation when filling in the puzzle.

CLUES

Across

2. Portable tanks, which usually hold a few hundred gallons (several hundred liters) of product and are characterized by a unique style of construction.
4. Diamond-shaped signage required to be placed on all four sides of highway transport vehicles, railroad tank cars, and other forms of hazardous materials transportation that identifies the hazardous contents of the vehicle.
5. One or more small openings in closed-head drums.
7. A reference book, written in plain language, to guide emergency responders in their initial actions at the incident scene.
8. A portable compressed gas container.
11. Any vessel or receptacle that holds material, including storage vessels, pipelines, and packaging.
14. Shipping papers for roads and highways.
15. Glass, plastic, or steel containers, ranging in volume from 5 to 15 gallons (19 to 57 liters).

Down

1. Shipping papers for trains.
3. High-volume transportation devices made up of several individual compressed gas cylinders banded together and affixed to a trailer.
4. A length of pipe including pumps, valves, flanges, control devices, strainers, and/or similar equipment for conveying fluids.
6. An agency maintained and staffed by the U.S. Coast Guard; it should always be notified if any spilled material could possibly enter a navigable waterway.
9. Shipping papers on an airplane.
10. Gaseous substances that have been chilled to the point at which they liquefy; a liquid having a boiling point lower than –150°F (–101°C) at 14.7 psia (an absolute pressure of 101 kPa).
11. A national call center for basic chemical information.
12. Diamond-shaped markings that are smaller versions of placards, placed on four sides of individual boxes and smaller packages.
13. A color-coded marking system by which employers give their personnel the necessary information to work safely around chemicals.

Fire Alarms

The following real case scenarios will give you an opportunity to explore the concerns associated with hazardous materials. Read each scenario, and then answer each question in detail.

1. Your engine company has responded to an explosion at a local abortion clinic. You are instructed to be aware that a secondary device might potentially be present.

    A. What is a secondary device?

    _____
    _____
    _____
    _____
    _____
    _____

    B. What are indicators of a potential secondary device?

    _____
    _____
    _____
    _____
    _____
    _____

2. Your engine company is first on the scene of a vehicle accident on the four-lane highway leading into town. You see an MC-307 lying on its side with liquid leaking from the valving. Your company officer orders you to gather information for a call to CHEMTREC.

    A. What information will you collect for the phone call?

    _____
    _____
    _____
    _____
    _____
    _____

    B. What is the CHEMTREC emergency phone number?

    _____
    _____
    _____
    _____
    _____
    _____

Fire Fighter II in Action

The following scenario will give you an opportunity to apply your firefighting knowledge and your fire department SOGs to the new information you learned while studying this chapter. Research your department's SOGs and answer the assignment in detail. Compare your answers with your classmates' and discuss similarities and obvious differences between your answers.

You are dispatched to an emergency medical call. The dispatch information given is a man down in a residential storage shed. Upon arrival, your crew walks to the back yard, noticing a small storage shed, a lawn mower beside the shed, and a swimming pool behind the house. As you approach the shed, you get a very strong smell of chlorine.

1. What would your initial size-up be?

   _____
   _____
   _____
   _____
   _____

2. What, if any, additional help would you request?

   _____
   _____
   _____
   _____
   _____

You look into the shed from a distance and see a man down on the shed floor. He is not moving.

3. What should you do to assist the victim?

   _____
   _____
   _____
   _____
   _____

Hazardous Materials: Implementing a Response

Workbook Activities

The following activities have been designed to help you. Your instructor may require you to complete some or all of these activities as a regular part of your fire fighter training program. You are encouraged to complete any activity that your instructor does not assign as a way to enhance your learning in the classroom.

Chapter Review

The following exercises provide an opportunity to refresh your knowledge of this chapter.

Matching

Match each of the terms in the left column to the appropriate definition in the right column.

_____ 1. Level I
_____ 2. Level III
_____ 3. Material safety data sheet (MSDS)
_____ 4. Defensive objectives
_____ 5. Level II

A. Highest level of threat
B. Actions that do not involve the actual stopping of the leak or release of a hazardous material
C. Level at which a hazardous materials response is needed
D. Usually states the concentration of the hazardous material and can be used to estimate the concentration of a chemical release
E. Lowest level of threat

Multiple Choice

Read each item carefully, and then select the best response.

_____ 1. When/where do secondary attacks take place?
   A. As responders treat victims
   B. At the firehouse
   C. At the police station
   D. Never

_____ 2. Defensive actions include
   A. plugging.
   B. patching.
   C. overpacking.
   D. diking.

_____ 3. When planning an initial hazardous materials incident response, what is the first priority?
   A. Consider the effect on the environment.
   B. Consider the safety of the victims.
   C. Consider the equipment and personnel needed to mediate the incident.
   D. Consider the safety of the responding personnel.

CHAPTER 31

_____ **4.** Victims removed from contaminated zones must be
   A. searched.
   B. confined.
   C. decontaminated.
   D. arrested.

_____ **5.** Planning a response begins with the
   A. size-up.
   B. initial call for help.
   C. incident commander's orders.
   D. review of standard operating procedures.

_____ **6.** Litmus paper is used to determine the
   A. time at which the contamination occurred.
   B. pH.
   C. weather.
   D. location of the contamination.

_____ **7.** Responders to a hazardous materials incident need to know the
   A. type of material involved.
   B. general operating guidelines.
   C. short- and long-term effects of the hazardous material.
   D. duration of the incident.

_____ **8.** The determination of which personal protective equipment is needed is based on the
   A. hazardous material involved.
   B. level of training of the responder.
   C. direction of the incident commander.
   D. standard operating procedures of the department.

_____ **9.** Responders to hazardous materials incidents need to consider
   A. the size of the container.
   B. the nature and amount of the material released.
   C. the area exposed to the material.
   D. all of the above.

_____ **10.** What is the main hub of the incident management system?
   A. The hot zone
   B. The command post
   C. The staging area
   D. The logistics tent

Vocabulary

Define the following terms using the space provided.

   1. Decontamination team:

   _____

   _____

   _____

2. Hazardous materials safety officer:
_____
_____
_____

3. Defensive objectives:
_____
_____
_____

4. Backup entry team:
_____
_____
_____

5. Hot zone entry team:
_____
_____
_____

Fill-in
Read each item carefully, and then complete the statement by filling in the missing word(s).

1. A primary terrorist attack may purposely injure members of the public to draw _____ into the scene.

2. At the operational level of training, all response objectives should be primarily _____ in purpose.

3. Monitoring devices such as wind direction and weather forecasting equipment are critical resources for the _____ _____ in formulating response plans.

4. When a hazardous materials incident is detected, there should be an initial call for additional _____.

5. The _____ of the affected area near the location of the spill or leak are important factors in planning the response to an incident.

6. When choosing a site for the ICP, the _____ margin of safety must be used.

7. The methods of decontamination are dictated by the _____ _____.

8. During a hazardous materials incident, no _____ action should be taken until the identity of the hazardous material involved is confirmed.

9. The basic incident management system consists of five sections: _____, _____, _____, _____, and _____.

10. _____ paper can be used to determine the concentration of an acid or a base by reporting the hazardous material's pH.

True/False

If you believe the statement to be more true than false, write the letter "T" in the space provided. If you believe the statement to be more false than true, write the letter "F."

_____ 1. If information regarding the hazardous material is unknown or is unconfirmed, the responders should prepare and approach the incident assuming that it involves the normal hazardous materials present in the area.

_____ 2. A predetermined list of contact names, agencies, and numbers should be established and maintained by each and every fire fighter.

_____ 3. The safety of responders is paramount to maintaining an effective response to any hazardous materials incident.

_____ 4. When dealing with a hazardous material, a variety of sources of information should be compared for consistency.

_____ 5. Monitoring and portable detection devices assist the incident commander in determining the hot, warm, and cold zones and the evacuation distances required.

Short Answer

Complete this section with short written answers using the space provided.

1. List the special technical groups that may develop under the Operations Section during an incident involving hazardous materials.

   _____
   _____
   _____
   _____
   _____

2. Identify and briefly describe the three hazardous materials incident levels.

   _____
   _____
   _____
   _____
   _____

3. Identify the three defensive objectives.

_____
_____
_____
_____
_____

4. Identify 11 pieces of information that could be reported to agencies to assist in their preparation for a response to a hazardous materials incident.

_____
_____
_____
_____
_____
_____
_____
_____
_____
_____
_____

Word Fun

The following crossword puzzle is an activity provided to reinforce correct spelling and understanding of terminology associated with firefighting. Use the clues provided to complete the puzzle. Do not include spaces or punctuation when filling in the puzzle.

CLUES

Across

4. The hazardous materials _____ is responsible for directing and coordinating all operations involving hazardous materials/weapons of mass destruction as assigned by the incident commander. (NFPA 472)
5. _____ objectives are actions that do not involve the actual stopping of the leak or release of a hazardous material, or contact of responders with the material; these include preventing further injury and controlling or containing the spread of the hazardous material.
6. The _____ team comprises fire fighters assigned to the entry into the designated hot zone.

Down

1. The _____ team is responsible for reducing and preventing the spread of contaminants from persons and equipment used at a hazardous materials incident.
2. The hazardous materials _____ works within the Incident Management System (specifically, the Hazardous Materials Branch/Group) to ensure that recognized hazardous materials/weapons of mass destruction (WMD) safe practices are followed at hazardous materials/WMD incidents. (NFPA 472)
3. A dedicated team of fully qualified and equipped responders who are ready to enter the hot zone at a moment's notice to rescue any member of the hot zone entry team.

Fire Alarms

The following real case scenario will give you an opportunity to explore the concerns associated with hazardous materials. Read the scenario, and then answer the question in detail.

1. You are on the scene of a large hazardous material release. Multiple agencies are en route to the site, and it appears that the incident will entail a lengthy response. Recognizing the severity of the incident, the incident commander tells your engine company to secure a location at which to establish a formal command post. How should you proceed?

___
___
___
___
___
___

Fire Fighter II in Action

The following scenario will give you an opportunity to apply your firefighting knowledge and your fire department SOGs to the new information you learned while studying this chapter. Research your department's SOGs and answer the assignment in detail. Compare your answers with your classmates' and discuss similarities and obvious differences between your answers.

Your crew has been dispatched to a vehicle accident on the interstate. As you arrive, you notice a tractor-trailer rig involved in an accident with an automobile. The truck is not placarded, but you notice a sizable fuel leak from the saddle tanks.

1. What are your initial concerns with this accident? What initial actions should you take to address these concerns?

2. As an operations level trained responder, what can you do to minimize the impact of the fuel release?

Hazardous Materials: Personal Protective Equipment, Scene Safety, and Scene Control

Workbook Activities

The following activities have been designed to help you. Your instructor may require you to complete some or all of these activities as a regular part of your fire fighter training program. You are encouraged to complete any activity that your instructor does not assign as a way to enhance your learning in the classroom.

Chapter Review

The following exercises provide an opportunity to refresh your knowledge of this chapter.

Matching
Match each of the terms in the left column to the appropriate definition in the right column.

_____ 1. Degradation
_____ 2. Level A
_____ 3. Cold zone
_____ 4. Level C
_____ 5. TLV/C
_____ 6. SARs
_____ 7. Hot zone
_____ 8. TLV/TWA
_____ 9. APRs
_____ 10. Warm zone

A. The safe area that houses the command post at an incident
B. The physical destruction of clothing following chemical exposure
C. The maximum concentration of hazardous material to which a worker should not be exposed, even for an instant
D. The area immediately surrounding a hazardous materials incident
E. The airborne concentration of a material to which a worker can be exposed for 8 hours a day, 40 hours a week, and not suffer any ill effects
F. Location of the decontamination corridor
G. Devices worn to filter particulates and contaminants from the air
H. Useful during extended operations such as decontamination, clean-up, and remedial work
I. Used when the hazardous material identified requires the highest level of protection for skin, eyes, and respiration
J. Worn when the airborne substance is known, criteria for APR are met, and skin and eye exposure is unlikely

Multiple Choice
Read each item carefully, and then select the best response.

_____ 1. One of the primary objectives of a medical surveillance program is to determine
   A. the intensity of the response at an incident.
   B. the concentration of the chemicals at an incident.
   C. the time of duty at an incident.
   D. any changes in the functioning of body systems.

CHAPTER 32

_____ 2. What is the safe area in which personnel do not need to wear any special protective clothing for safe operation called?
   A. Control zone
   B. Hot zone
   C. Warm zone
   D. Cold zone

_____ 3. Chemical-protective clothing is rated for its effectiveness against chemical permeation, including how quickly it protects the fire fighter and
   A. how well it fits the fire fighter.
   B. how many times the suit can be used.
   C. to what degree it protects the fire fighter.
   D. how visible the fire fighter is when wearing the suit.

_____ 4. What is the area where personnel and equipment are staged before they enter and after they leave the hot zone called?
   A. Control zone
   B. Hot zone
   C. Warm zone
   D. Cold zone

_____ 5. After team members undergo decontamination, they should
   A. prepare for reassignment.
   B. remove all layers of their protective uniforms.
   C. have all vital signs checked.
   D. report to the incident commander.

_____ 6. The level of PPE required for responding to a hazardous materials incident should be approved by the
   A. incident commander.
   B. safety officer.
   C. hazardous material technician.
   D. crew captain.

_____ 7. Designated areas at a hazardous materials incident based on safety and the degree of hazard are called
   A. control zones.
   B. hot zones.
   C. warm zones.
   D. cold zones.

_____ 8. The process by which a hazardous chemical moves through closures, seams, or porous materials is called
   A. penetration.
   B. degradation.
   C. permeation.
   D. vaporization.

_____ 9. The first step in gaining control of a hazardous materials incident is to isolate the problem and
   A. equip the cold zone.
   B. keep people away.
   C. establish a backup team.
   D. identify the hazardous materials involved.

_____ 10. What is the area immediately around and adjacent to the incident called?
   A. Control zone
   B. Hot zone
   C. Warm zone
   D. Cold zone

_____ 11. The physical destruction of clothing due to chemical exposure is called
   A. penetration.
   B. degradation.
   C. permeation.
   D. vaporization.

_____ 12. Chemical resistance, flexibility, abrasion, temperature resistance, shelf life, and sizing criteria are requirements that need to be considered when selecting
   A. entry tools.
   B. respirators.
   C. testing equipment.
   D. chemical-protective materials.

_____ 13. If a person's body temperature falls below 95°F (35°C), he or she may experience
   A. hypothermia.
   B. death.
   C. hyperthermia.
   D. cold exhaustion.

_____ 14. Air-purifying respirators should be worn in atmospheres where the type and quantity of contaminants are
   A. unknown.
   B. known.
   C. suspected.
   D. indistinguishable.

_____ 15. The principal dangers of hazardous materials are toxicity, flammability, and
   A. reactivity.
   B. instability.
   C. tolerance.
   D. transportability.

Vocabulary

Define the following terms using the space provided.

1. Backup personnel:

   _____
   _____
   _____

2. Immediately dangerous to life and health (IDLH):

   _____
   _____
   _____

3. High temperature-protective clothing:

   _____
   _____
   _____

4. Heat stroke:

5. Heat exhaustion:

Fill-in
Read each item carefully, and then complete the statement by filling in the missing word(s).

1. _____-protective clothing is designed to prevent chemicals from coming in contact with the body and may have varying degrees of resistance.

2. _____ _____ is enhanced by abrasions, cuts, heat, and moisture.

3. Fire fighters should be encouraged to drink _____ to _____ ounces of water before donning any protective clothing.

4. The layer of clothing next to the skin, especially the _____ should always be kept dry.

5. Work uniforms offer the _____ amount of protection in a hazardous materials emergency.

6. _____ are most likely to penetrate material.

7. The high absorbency rate of the _____ makes them more susceptible than normal skin to contaminants and one of the fastest means of exposure.

8. Wet clothing extracts heat from the body as many as _____ times faster than dry clothing.

9. An encapsulated suit is a(n) _____-piece garment that completely encloses the wearer.

10. _____ burns are often much deeper and more destructive than acid burns.

True/False

If you believe the statement to be more true than false, write the letter "T" in the space provided. If you believe the statement to be more false than true, write the letter "F."

_____ 1. The backup personnel remain on standby in the cold zone awaiting orders to prepare for follow-up duties.

_____ 2. All personnel must be fully briefed before they approach the hazard area or enter the cold zone.

_____ 3. All members of the responding team must know the shielding capabilities and limitations of their personal protective clothing.

_____ 4. An incident that involves a gaseous contaminant will require a larger hot zone than one involving a liquid leak.

_____ 5. There are several ways to isolate the hazard area and create the control zones.

_____ 6. The warm zone contains control points for access corridors as well as the decontamination corridor.

_____ 7. A hazardous materials incident may require different levels of PPE.

_____ 8. Tyvek provides satisfactory protection from all chemicals.

_____ 9. When possible, approach a hazardous materials incident cautiously from downwind of the site.

_____ 10. The decontamination team must be in place before anyone enters the hot zone.

Short Answer

Complete this section with short written answers using the space provided.

1. Identify and provide a brief description of the three zones at a hazardous materials incident.

   _____
   _____
   _____
   _____
   _____

2. Identify and define the three basic atmospheres at a hazardous materials emergency according to the exposure guidelines.

   _____
   _____
   _____
   _____
   _____

3. Identify and provide a brief description of the four levels of protective clothing.

   _____
   _____
   _____
   _____
   _____

Word Fun

The following crossword puzzle is an activity provided to reinforce correct spelling and understanding of terminology associated with firefighting. Use the clues provided to complete the puzzle. Do not include spaces or punctuation when filling in the puzzle.

CLUES

Across

1. Painful muscle spasms usually associated with vigorous activity in a hot environment.
4. The control zone at hazardous materials/weapons of mass destruction incidents where personnel and equipment decontamination and hot zone support take place. (NFPA 472)
7. The movement of a material through a suit's closures, such as zippers, buttonholes, seams, flaps, or other design features of chemical-protective clothing, and through punctures, cuts, and tears. (NFPA 472)
9. Clear, raised bumps on the skin caused by friction and sweat.
10. A chemical action involving the molecular breakdown of a protective clothing material or equipment caused by contact with a chemical. (NFPA 472)
11. An area at hazardous materials/weapons of mass destruction incidents within an established perimeter that is designated based upon safety and the degree of hazard. (NFPA 472)

Down

1. The control zone immediately surrounding hazardous materials/weapons of mass destruction incidents, which extends far enough to prevent adverse effects of hazards to personnel outside the zone. (NFPA 472)
2. Air-purifying respirator with a hood or helmet, breathing tube, canister, cartridge, filter, and a blower that passes ambient air through the purifying element. (NFPA 1404)
3. An atmosphere-supplying respirator for which the source of breathing air is not designed to be carried by the user. (NFPA 1404)
5. A severe and sometimes fatal condition resulting from the failure of the temperature-regulating capacity of the body. It is caused by prolonged exposure to the sun or high temperatures.
6. A chemical action involving the movement of chemicals, on a molecular level, through intact material. (NFPA 472)
7. The maximum permitted eight-hour, time-weighted average concentration of an airborne contaminant. (NFPA 5000)
8. The control zone of hazardous materials/weapons of mass destruction incidents that contains the incident command post and such other support functions as are deemed necessary to control the incident. (NFPA 472)
12. The established standard limit of exposure to a hazardous material. It is based on the maximum time-weighted concentration at which 95 percent of exposed, healthy adults suffer no adverse effects over a 40-hour workweek.

Fire Alarms

The following real case scenarios will give you an opportunity to explore the concerns associated with hazardous materials. Read each scenario, and then answer each question in detail.

1. After 45 minutes of training while wearing your Level B PPE, you begin feeling dizzy and are sweating profusely. You are also feeling weak and notice some blurring of your vision.

   A. What is the probable cause of your sudden illness?

   B. Which actions should be taken?

2. During hazardous materials response training, you are assigned to an entry team wearing Level B nonencapsulated personal protective clothing. What is the recommended PPE for Level B protection?

Fire Fighter II in Action

The following scenarios will give you an opportunity to apply your firefighting knowledge and your fire department SOGs to the new information you learned while studying this chapter. Research your department's SOGs and answer the assignments in detail. Compare your answers with your classmates' and discuss similarities and obvious differences between your answers.

1. Chemical-protective clothing has many limitations. List six of these limitations, and describe how you will work within the limitations.

2. One limitation for Level A and Level B clothing is the need for a breathing air supply. Discuss air management issues and ideas to overcome these issues.

Skill Drills

Skill Drill 32-1: Donning a Level B Encapsulated Chemical-Protective Clothing Ensemble
Test your knowledge of this skill drill by filling in the correct words in the photo captions.

1. While seated, pull on the suit to _____ level; put on the _____ _____ over the chemical suit. Pull the suit _____ _____ over the tops of the boots.

2. Stand up and don SCBA and the SCBA _____ _____, but do not connect the _____ to the face piece.

3. Place the _____ on the head, if required.

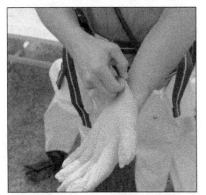

4. Don the _____ _____.

5. With assistance, complete donning the suit. Instruct the assistant to connect the _____ to the SCBA _____ _____ and ensure _____ _____.

6. Instruct the assistant to close the chemical suit by closing the zipper and sealing the _____ _____.

7. Review _____ _____ and indicate that you are ready to operate.

Skill Drill 32-2: Doffing a Level B Encapsulated Chemical-Protective Clothing Ensemble
Test your knowledge of this skill drill by filling in the correct words in the photo captions.

© Jones & Bartlett Learning. Photographed by Glen E. Ellman.

1. After completing decontamination, proceed to the _____ area. Remove the _____ and _____ from the suit gloves and sleeves, and cross the arms in front inside the suit.

© Jones & Bartlett Learning. Photographed by Glen E. Ellman.

2. Instruct the assistant to open the chemical _____ _____ and open suit zipper.

© Jones & Bartlett Learning. Photographed by Glen E. Ellman.

3. Instruct the assistant to begin at the head and roll the suit down and away from you until the suit is below _____ level.

© Jones & Bartlett Learning. Photographed by Glen E. Ellman.

4. Sit and instruct the assistant to complete rolling down the suit and remove the _____ _____ and suit. Rotate on the bench to the direction that will allow you to place feet on a dry, clean area.

© Jones & Bartlett Learning. Photographed by Glen E. Ellman.

5. Stand and doff the SCBA using the _____-_____ method. Keep the _____ in place while the SCBA frame is placed on the ground.

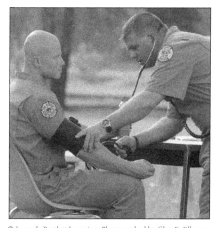

© Jones & Bartlett Learning. Photographed by Glen E. Ellman.

6. Take a deep breath, doff the SCBA mask, and walk away from the clean area. Go to the _____ area for _____ monitoring, rehydration, and personal decontamination shower.

Hazardous Materials: Response Priorities and Actions

Workbook Activities

The following activities have been designed to help you. Your instructor may require you to complete some or all of these activities as a regular part of your fire fighter training program. You are encouraged to complete any activity that your instructor does not assign as a way to enhance your learning in the classroom.

Chapter Review

The following exercises provide an opportunity to refresh your knowledge of this chapter.

Matching

Match each of the terms in the left column to the appropriate definition in the right column.

_____ 1. Absorption
_____ 2. Containment
_____ 3. Vapor dispersion
_____ 4. Damming
_____ 5. Confinement
_____ 6. Diking
_____ 7. Diversion
_____ 8. Dilution
_____ 9. Retention
_____ 10. Vapor suppression

A. The process of adding some substance to a product to weaken its concentration
B. The process of lowering the concentration of vapors by spreading them out
C. The process of applying a material that will soak up and hold the hazardous material
D. Used when a liquid is flowing in a natural channel or depression and its progress can be stopped by blocking the channel
E. The process of keeping a hazardous material on the site or within the immediate area of the release
F. Actions relating to stopping a hazardous materials container from leaking, such as patching, plugging, or righting the container
G. The placement of impervious materials to form a barrier that will keep a hazardous material in liquid form from entering an area
H. Redirecting spilled material to an area where it will have less impact
I. The process of creating an area to hold hazardous materials
J. The process of controlling vapors by covering the product with foam or by reducing the temperature of the material

CHAPTER 33

Multiple Choice
Read each item carefully, and then select the best response.

_____ 1. Why is the technique of absorption difficult for operational personnel to implement?
   A. It creates an extensive clean-up process.
   B. It requires the appropriate material matches.
   C. It involves a large number of personnel.
   D. It generally involves being in close proximity to the spill.

_____ 2. Which of the following is one of the first response priorities at a hazardous materials incident?
   A. Contacting the property owners
   B. Evacuating the exposed area
   C. Containing the hazardous materials
   D. Alerting the appropriate responding agencies

_____ 3. The process of lowering the concentration of vapors by spreading them out is called
   A. vapor suppression.
   B. vapor release.
   C. vapor evacuation.
   D. vapor dispersion.

_____ 4. Who makes the decision to terminate a hazardous materials incident?
   A. Safety officer
   B. Incident commander
   C. Operations officer
   D. Planning officer

_____ 5. When the imminent danger has passed and clean-up and the return to normalcy have begun, the incident has reached the
   A. debriefing phase.
   B. clean-up phase.
   C. recovery phase.
   D. termination phase.

_____ 6. To determine how far to extend evacuation distances, hazardous materials technicians should
   A. contact product specialists.
   B. refer to the ERG.
   C. get direction from the incident commander.
   D. use detection and monitoring equipment.

_____ 7. Which phase of the incident includes the compilation of all records necessary for documentation of the incident?
   A. Administration phase
   B. Recovery phase
   C. Wrap-up phase
   D. Size-up phase

_____ 8. The process of attempting to keep the hazardous material on the site or within the immediate area of the release is known as
   A. confinement.
   B. containment.
   C. exposure.
   D. suppression.

_____ 9. The addition of another liquid to weaken the concentration of a hazardous material is called
   A. dispersion.
   B. dilution.
   C. extension.
   D. liquidation.

_____ 10. Which of the following factors is a major concern when considering evacuation?
   A. Potential for exposure to the material
   B. Distance to a safe area
   C. Time of day
   D. Amount of property involved in the incident

_____ 11. Most flammable and combustible liquid fires can be extinguished by the use of
   A. water.
   B. carbon monoxide.
   C. foam.
   D. dilution.

_____ 12. Which method of safeguarding people in a hazardous area involves keeping them in a safe atmosphere?
   A. Staying indoors
   B. Duck and cover
   C. Shelter-in-place
   D. Containment

_____ 13. The process of creating an area to hold hazardous materials is called
   A. retention.
   B. diking.
   C. damming.
   D. diversion.

_____ 14. As fire fighters approach a hazardous materials incident, they should look for
   A. a means of egress.
   B. damage to property or surfaces.
   C. natural control points.
   D. plumes of smoke.

_____ 15. During the initial size-up at a hazardous materials incident, the first decision concerns
   A. the amount of property affected.
   B. personnel safety.
   C. the number of people involved.
   D. the type of material involved.

Vocabulary

Define the following terms using the space provided.

1. Shelter-in-place:

   _____
   _____
   _____

2. Recovery phase:

   _____
   _____
   _____

3. Exposures (hazardous materials):

   _____
   _____
   _____

Fill-in

Read each item carefully, and then complete the statement by filling in the missing word(s).

1. Victim search is _____ _____ if a hazardous material exposure is not survivable.
2. Only a responder at the _____ level would patch or plug a container.
3. Most flammable and combustible _____ _____ can be extinguished by the use of foam.
4. _____ is the process of creating an area to hold hazardous materials.
5. Before an evacuation order is given, a(n) _____ _____ and suitable shelter are established.
6. The protection of _____ is the first priority in any emergency response situation.
7. Incidents involving pressurized-gas _____ may involve fires and/or releases of their contents.
8. The _____ of the hazardous material is a major factor in the decision whether to evacuate.
9. If fire fighters expose themselves to _____ risk, injury, exposure, or contamination, they only complicate the problem.
10. A(n) _____ _____ is placed across a small stream or ditch to completely stop the flow of materials through the channel.

Fundamentals of Fire Fighter Skills

True/False

If you believe the statement to be more true than false, write the letter "T" in the space provided. If you believe the statement to be more false than true, write the letter "F."

_____ 1. In some cases. the incident commander may decide to withdraw to a safe distance and let the hazardous materials incident run its course.

_____ 2. In a hazardous materials incident, all emergency response personnel must first recognize and identify which hazardous materials may be present.

_____ 3. A retention technique is used to redirect the flow of a liquid away from an area.

_____ 4. MC 307/DOT 407 cargo tanks are certified to carry chemicals that are transported at high pressure.

_____ 5. Dilution can be used only when the identity and properties of the hazardous material are known with certainty.

_____ 6. The duration of the hazardous materials incident is a factor in determining whether shelter-in-place is a viable option.

_____ 7. Many chemical processes, or piped systems that carry chemicals, have a way to shut down a system or isolate a valve remotely.

_____ 8. All exposures need to be protected in the same way.

_____ 9. The recovery phase and clean-up will likely require amounts of resources and equipment that are far beyond the capabilities of local responders.

_____ 10. Firefighting foams should be sprayed directly on the burning material and surface.

Short Answer

Complete this section with a short written answer using the space provided.

1. List three of the types of firefighting foams, and explain what each is used for.

_____
_____
_____
_____
_____
_____
_____
_____
_____
_____

Word Fun

The following crossword puzzle is an activity provided to reinforce correct spelling and understanding of terminology associated with firefighting. Use the clues provided to complete the puzzle. Do not include spaces or punctuation when filling in the puzzle.

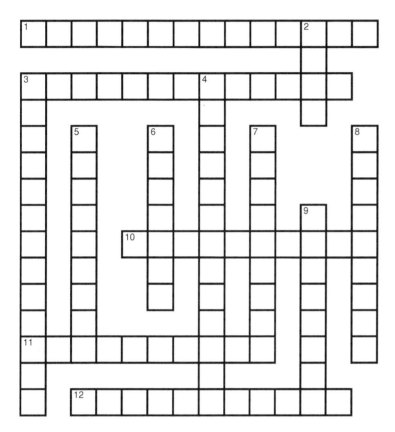

CLUES

Across

1. A method of safeguarding people in a hazardous area by keeping them in a safe atmosphere, usually inside structures.
3. Valve device designed to retain a hazardous material in its container.
10. The process of lowering the concentration of vapors by spreading them out.
11. The process of applying a material that will soak up and hold a hazardous material in a sponge-like manner, for collection and disposal.
12. Actions taken to keep a material, once released, in a defined or local area.

Down

2. Type of foam that is designed to form a blanket over spilled flammable liquids to suppress vapors, or on actively burning pools of flammable liquids.
3. The stage of a hazardous materials incident after the imminent danger has passed, when clean-up and the return to normalcy have begun.
4. Type of foam used when large volumes of foam are required for spills or fires in warehouses, tank farms, and hazardous waste facilities.
5. Consists of people, property, structures, or the environment that are subject to influence, damage, or injury as a result of contact with the hazardous material.
6. A process used when liquid is flowing in a natural channel or depression and its progress can be stopped by blocking the channel.
7. Redirecting spilled or leaking material to an area where it will have less impact.
8. The process of purposefully collecting hazardous materials in a defined area. It may include creating that area by digging, for instance.
9. The process of adding some substance, usually water, to a product to weaken its concentration.

Fundamentals of Fire Fighter Skills

Fire Alarms

The following real case scenarios will give you an opportunity to explore the concerns associated with hazardous materials. Read each scenario, and then answer each question in detail.

You have been dispatched to a diesel fuel spill on the state route south of town. On arrival at the site, you find 10 gallons of diesel fuel spilled across the roadway. Your company officer directs you to absorb the fuel with on-board absorbent.

1. How will you absorb the spilled hazardous material?

   _____
   _____
   _____
   _____
   _____
   _____

Before you can begin absorbing the spilled diesel fuel, the combustible material is ignited by a road flare placed too close to the spill.

2. What can you use to extinguish the burning combustible fuel?

   _____
   _____
   _____
   _____
   _____
   _____

3. How should the extinguishing agent be applied?

Fire Fighter II in Action

The following scenario will give you an opportunity to apply your firefighting knowledge and your fire department SOGs to the new information you learned while studying this chapter. Research your department's SOGs and answer the assignment in detail. Compare your answers with your classmates' and discuss similarities and obvious differences between your answers.

All of your combustible gas indicators have been calibrated to propane gas.

1. Can these meters be used on a methane gas release?

2. How can you determine their effectiveness?

_____
_____
_____
_____
_____

3. What factors will limit the effectiveness of foam as an extinguishing agent on a large, very hot flammable liquid fire?

_____
_____
_____
_____
_____

Skill Drills

Skill Drill 33-2: Using Absorption/Adsorption to Manage a Hazardous Materials Incident
Test your knowledge of this skill drill by filling in the correct word(s) in the photo captions.

1. Decide which _____ is best suited with the spilled product. Assess location of spill and stay clear of _____ _____.

2. Apply appropriate material to _____ and _____ the spilled material.

3. Maintain materials and take appropriate steps for their _____.

Hazardous Materials: Decontamination Techniques

Workbook Activities

The following activities have been designed to help you. Your instructor may require you to complete some or all of these activities as a regular part of your fire fighter training program. You are encouraged to complete any activity that your instructor does not assign as a way to enhance your learning in the classroom.

Chapter Review

The following exercises provide an opportunity to refresh your knowledge of this chapter.

Matching

Match each of the terms in the left column to the appropriate definition in the right column.

_____ 1. Emulsification

_____ 2. Adsorption

_____ 3. Decontamination

_____ 4. Vacuuming

_____ 5. Disinfection

_____ 6. Vapor dispersion

_____ 7. Removal

_____ 8. Solidification

_____ 9. Dilution

_____ 10. Absorption

A. The process of chemically treating a hazardous liquid to turn it into a solid material, making the material easier to handle

B. The process of removing any form of contaminant from a person, an object, or the environment

C. The removal of dusts, particles, and some liquids by sucking them up into a container

D. The process of adding a material to a contaminant, which then adheres to the surface of the material for collection

E. The process used to destroy recognized pathogenic microorganisms

F. The process of changing the chemical properties of a hazardous material, thereby reducing its harmful effects

G. The process of separating and diminishing harmful vapors

H. The process of mixing a spongy material into a spilled liquid and picking up the mixture together

I. A mode of decontamination that applies specifically to contaminated soil that is taken away from the scene

J. Uses plain water or a soap-and-water mixture to lower the concentration of a hazardous material while flushing it off a contaminated person or object

Multiple Choice

Read each item carefully, and then select the best response.

_____ 1. Hazardous materials that have been emulsified should be
   A. diluted.
   B. solidified.
   C. bagged and tagged.
   D. disposed of properly.

CHAPTER 34

_____ 2. Which decontamination procedure mixes a spongy material with a liquid hazardous material?
   A. Absorption
   B. Adsorption
   C. Dilution
   D. Vapor dispersal

_____ 3. What is the term for when a contaminated person comes into direct contact with another person or object?
   A. Cross-contamination
   B. Dispersion
   C. Transference
   D. Integration

_____ 4. All personal clothing should be
   A. diluted.
   B. solidified.
   C. bagged and tagged.
   D. burned.

_____ 5. Who is responsible for laws governing the disposal of absorbent materials?
   A. Fire department
   B. Government
   C. Department of Transportation
   D. Emergency response team

_____ 6. After personnel are thoroughly decontaminated, they should proceed to
   A. the rehabilitation area.
   B. EMS personnel.
   C. the incident commander.
   D. the Operations Section.

_____ 7. Which method of decontamination is used during incidents involving unknown agents and large groups of people?
   A. Emergency decontamination
   B. Group decontamination
   C. Gross decontamination
   D. Mass decontamination

_____ 8. Which of the following is a two-step removal process for items that cannot be properly decontaminated?
   A. Disinfection
   B. Solidification
   C. Disposal
   D. Rapid mass decontamination

_____ 9. Removed equipment should be placed
   A. on the contaminated side of the corridor.
   B. in the hot zone.
   C. in the cold zone.
   D. in the hazardous materials truck.

_____ 10. During decontamination, what is usually the last item of clothing removed?
   A. Shoes
   B. SCBA mask
   C. Inner gloves
   D. Face shield

Vocabulary
Define the following terms using the space provided.

1. Adsorption:
_____
_____
_____

2. Decontamination team:
_____
_____
_____

3. Emulsification:
_____
_____
_____

4. Contamination:
_____
_____
_____

5. Solidification:
_____
_____
_____

Fill-in
Read each item carefully, and then complete the statement by filling in the missing word(s).

1. Whenever possible, _____ the hazardous material before beginning decontamination.

2. The opposite of absorption is _____.

3. The _____ _____ is a controlled area, usually within the warm zone, where decontamination procedures take place.

4. The process of separating and diminishing harmful vapors is known as _____ _____.

5. Do not allow the water runoff from emergency decontamination to flow into _____, _____, or _____.

6. The mode of decontamination that applies specifically to contaminated soil that can be taken away from the scene is called _____.

7. During gross decontamination, runoff water should be controlled because it is likely to contain _____.

8. Fire fighters tend to use _____ as the first decontamination method.

9. _____ _____ is performed after gross decontamination and is a more thorough cleaning process.

10. A water spray is commonly used to _____ vapors.

True/False

If you believe the statement to be more true than false, write the letter "T" in the space provided. If you believe the statement to be more false than true, write the letter "F."

_____ 1. Contact lenses can trap contaminants and therefore need to be removed during decontamination.

_____ 2. Personnel leaving the hot zone should place used tools in a tool drop area near the decontamination corridor.

_____ 3. During gross decontamination, hospital staff use low-pressure, high-volume water flow to rinse off and dilute contaminants.

_____ 4. Emergency medical responders are responsible for establishing a decontamination corridor for the initial emergency response crews and victims.

_____ 5. Vacuuming is the removal of dusts, particles, and some liquids by sucking them into a container.

Short Answer

Complete this section with a short written answer using the space provided.

1. Identify and provide a brief description of the four major categories of decontamination.

   _____
   _____
   _____
   _____
   _____
   _____

Clues

Across

2. The process used to inactivate virtually all recognized pathogenic microorganisms but not necessarily all microbial forms, such as bacterial endospores. (NFPA 1581)
3. The physical and/or chemical process of reducing and preventing the spread of contaminants from people, animals, the environment, or equipment involved at hazardous materials/weapons of mass destruction incidents. (NFPA 472)
6. Decontamination technique in which a spongy material is mixed with a liquid hazardous material. The contaminated mixture is collected and disposed of.
8. The process of changing the chemical properties of a hazardous material to reduce its harmful effects.

Down

1. The process of chemically treating a hazardous liquid so as to turn it into a solid material, thereby making the material easier to handle.
3. A two-step removal process for contaminated items that cannot be properly decontaminated. Items are bagged and placed in appropriate containers for transport to a hazardous waste facility.
4. The process of adding a material such as sand or activated carbon to a contaminant, which then adheres to the surface of the material. The contaminated material is then collected and disposed of.
5. The process of cleaning up dusts, particles, and some liquids using a vacuum with high-efficiency particulate air filtration to prevent recontamination of the environment.
7. A mode of decontamination that applies specifically to contaminated soil, which is taken away from the scene.

Word Fun

The following crossword puzzle is an activity provided to reinforce correct spelling and understanding of terminology associated with firefighting. Use the clues provided to complete the puzzle. Do not include spaces or punctuation when filling in the puzzle.

Fire Alarms

The following real case scenarios will give you an opportunity to explore the concerns associated with hazardous materials. Read each scenario, and then answer each question in detail.

1. Your Lieutenant has given you the chance to prepare a short presentation on alternative decontamination procedures. Which topics will you discuss?

2. Your engine company is dispatched to a pesticide spill at a local hardware store. On arrival, you see several store patrons covered with liquid and powder. Your officer orders your company to set up emergency decontamination for the contaminated patrons. How will you proceed?

Fire Fighter II in Action

The following scenario will give you an opportunity to apply your firefighting knowledge and your fire department SOGs to the new information you learned while studying this chapter. Research your department's SOGs and answer the assignment in detail. Compare your answers with your classmates' and discuss similarities and obvious differences between your answers.

Your Hazmat Ops Instructor has given you an assignment to sketch a decontamination corridor for a complete technical decontamination.

1. Using your department's decontamination SOPs and equipment, sketch a technical decontamination corridor. Include required PPE and all necessary equipment.

Skill Drills

Skill Drill 34-3: Performing Technical Decontamination

Test your knowledge of this skill drill by placing the photos below in the correct order. Number the first step with a "1," the second step with a "2," and so on.

_____ Perform gross decontamination.

_____ Shower and wash the body. Dry off the body and put on clean clothing. The entry team member should proceed to medical monitoring. At the station, the entry members should fill out any record-keeping information about the incident to include exposure reporting.

_____ Remove outer hazardous materials–protective clothing and isolate PPE.

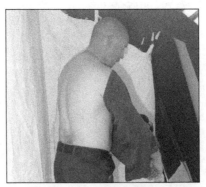

_____ Remove personal clothing. Bag and tag all personal clothing.

Chapter 34: Hazardous Materials: Decontamination Techniques

© Jones & Bartlett Learning. Photographed by Glen E. Ellman.

_____ Remove SCBA and the face piece.

© Jones & Bartlett Learning. Photographed by Glen E. Ellman.

_____ Drop any tools and equipment into a tool drum or onto a designated tarp.

© Jones & Bartlett Learning. Photographed by Glen E. Ellman.

_____ Wash and rinse the entry team member.

Terrorism Awareness

Workbook Activities

The following activities have been designed to help you. Your instructor may require you to complete some or all of these activities as a regular part of your fire fighter training program. You are encouraged to complete any activity that your instructor does not assign as a way to enhance your learning in the classroom.

Chapter Review

The following exercises provide an opportunity to refresh your knowledge of this chapter.

Matching

Match each of the terms in the left column to the appropriate definition in the right column.

_____ 1. Weapon of mass destruction (WMD)
_____ 2. Phosgene
_____ 3. Incubation period
_____ 4. Secondary device
_____ 5. Triage
_____ 6. Soman
_____ 7. Sarin
_____ 8. Decontamination
_____ 9. Chlorine
_____ 10. ANFO
_____ 11. Choking agent
_____ 12. Sulfur mustard
_____ 13. Nerve agent
_____ 14. Cyanide
_____ 15. Radiological agents

A. Materials that emit radioactivity
B. A chemical agent that causes severe pulmonary damage
C. A device that is designed to cause maximum damage to property or people
D. A chemical designed to inhibit breathing, which is typically designed to incapacitate rather than kill
E. Time period between the initial infection by an organism and the development of symptoms
F. An explosive device designed to injure emergency responders who have responded to an initial event
G. An explosive material containing ammonium nitrate fertilizer and fuel oil
H. A nerve agent that is primarily a vapor hazard
I. The process of sorting victims based on the severity of injury and medical needs to establish treatment and transportation priorities
J. A nerve gas that is both a contact and a vapor hazard that has the odor of camphor
K. A yellowish gas that has many industrial uses but also damages the lungs when inhaled
L. Toxic substances that attack the central nervous system in humans
M. A clear, yellow, or amber oily liquid with a faint sweet odor of mustard or garlic that may be dispersed in an aerosol form
N. The physical or chemical process of removing any form of contaminant from a person, an object, or the environment
O. A highly toxic chemical agent that attacks the circulatory system

CHAPTER 35

Multiple Choice

Read each item carefully, and then select the best response.

_____ 1. Which name is given to the time period between the actual infection and the appearance of symptoms?
   A. Growth period
   B. Dispersing period
   C. Incubation period
   D. Implementation period

_____ 2. A terrorist threat requires fire fighters to work closely with
   A. local, state, and federal law enforcement agencies.
   B. emergency management agencies.
   C. the military.
   D. all of the above.

_____ 3. What is the process of sorting victims based on the severity of their injuries and medical needs to establish treatment and transportation priorities called?
   A. EMS
   B. Decon
   C. Triage
   D. Beta

_____ 4. Fire fighters and emergency responders must remember that a terrorist incident is also a(n)
   A. crime scene.
   B. opportunity to improve working relations between departments.
   C. opportunity to implement advanced rescue techniques.
   D. all of the above.

_____ 5. Disrupting or deleting government or banking computer systems is an example of
   A. ecoterrorism.
   B. cyberterrorism.
   C. agroterrorism.
   D. religious terrorism.

_____ 6. For what purpose is a personal dosimeter used?
   A. To record personal exposure to contaminants
   B. To decrease personal exposure to contaminants
   C. To measure the amount of radioactive exposure
   D. To measure the active agents in the area

_____ 7. An IED is an explosive device that is contained in a package. IED is an acronym for
   A. improvised explosive device.
   B. internal explosive device.
   C. imploding explosive device.
   D. illuminating explosive device.

_____ 8. Fire fighters responding to a potential or known terrorist incident should use the same approach as they would when responding to a(n)
   A. structural fire.
   B. EMS incident.
   C. rescue incident.
   D. hazardous materials incident.

_____ 9. Bombing a store that sells fur coats would be an example of
   A. ecoterrorism.
   B. cyberterrorism.
   C. agroterrorism.
   D. religious terrorism.

_____ 10. Attacking a food industry or supply is an example of
   A. ecoterrorism.
   B. cyberterrorism.
   C. agroterrorism.
   D. religious terrorism.

_____ 11. Before anyone is allowed to enter a building involved in an explosion, what must happen?
   A. The utilities must be disconnected.
   B. All emergency response teams must arrive.
   C. Team members must review preincident plans.
   D. The stability of the building must be evaluated.

_____ 12. What are the three types of radiation?
   A. Internal, external, and dispersement
   B. Alpha particles, beta particles, and gamma rays
   C. Alpha, beta, and gamma particles
   D. Alpha particles, beta particles, and sigma rays

_____ 13. At an incident where there is potential terrorist or secondary device activity, the fire department should be part of a joint command structure commonly referred to as
   A. a team command.
   B. an emergency response team.
   C. a unified command.
   D. a united command.

_____ 14. Potential terrorist targets include
   A. bridges.
   B. pipelines.
   C. power plants.
   D. all of the above.

_____ 15. During a bomb disposal, where does the rapid intervention team stand by to provide immediate assistance?
   A. Bomb disposal containment area
   B. Forward staging area
   C. Incident command center
   D. Response area

Vocabulary

Define the following terms using the space provided.

1. Smallpox:

2. Radiation dispersal device:

3. Forward staging area:

4. V-agent:

5. Plague:

6. Universal precautions:

7. Tabun:

Fill-in

Read each item carefully, and then complete the statement by filling in the missing word(s).

1. Decontamination of a large number of victims or emergency responders is referred to as _____ _____.

2. If contamination is suspected, a plan must ensure that it does not spread beyond a(n) _____ _____.

3. _____ _____ release energy in the form of electromagnetic waves or energy particles that cannot be detected by the senses.

4. Decontamination should occur as soon as possible to prevent further _____ of a contaminant and to reduce the possibility of spreading the contamination.

5. The most common improvised explosive device is the _____ _____.

6. The time period between the actual infection and the appearance of symptoms is known as the _____ _____.

7. The most common dispersal method for chemical agents is _____ _____.

8. _____ can be described as the unlawful use of violence or threats of violence to intimidate or coerce a person or group to further political or social objectives.

9. _____ is a mnemonic used to remember the symptoms of possible nerve agent exposure.

10. _____ _____ are toxic substances used to attack the central nervous system and were first developed in Germany before World War II.

True/False

If you believe the statement to be more true than false, write the letter "T" in the space provided. If you believe the statement to be more false than true, write the letter "F."

_____ 1. Fire fighters must become familiar with potential terrorist targets and actions because they are often involved in the initial response and handling of a terrorist incident.

_____ 2. Anthrax and the plague are examples of nerve agents.

_____ 3. Emergency responders are decontaminated after they leave the contaminated area.

_____ 4. Exposure to high levels of radiation can cause vomiting and digestive system damage within a short time.

_____ 5. The first emergency response units to arrive should establish an outer perimeter to control access to and from the scene.

_____ 6. Beta particles are also active nerve agents.

_____ 7. A personal dosimeter is used to measure the amount of radioactive exposure.

_____ 8. Gamma rays are the least harmful of the three types of radiation.

_____ 9. Universal precautions must be enacted during acts of cyberterrorism.

_____ 10. Soman is a highly infectious disease that kills approximately 30 percent of those infected with it.

Short Answer

Complete this section with short written answers using the space provided.

1. Describe why responding to a terrorist incident puts fire fighters and emergency personnel at increased risk.

   _____
   _____
   _____
   _____
   _____

2. Identify the three types of radiation and ways to limit exposure to each.

   _____
   _____
   _____
   _____
   _____
   _____

3. Describe ecoterrorism, cyberterrorism, and agroterrorism.

   _____
   _____
   _____
   _____
   _____
   _____

4. Describe the issues that fire fighters must consider following a large explosion.

   _____
   _____
   _____
   _____
   _____

5. What motivates terrorists?

   _____
   _____
   _____
   _____
   _____

Word Fun

The following crossword puzzle is an activity provided to reinforce correct spelling and understanding of terminology associated with firefighting. Use the clues provided to complete the puzzle. Do not include spaces or punctuation when filling in the puzzle.

CLUES

Across

1. A nerve gas that is both a contact hazard and a vapor hazard; it has the odor of camphor.
5. An explosive made of commonly available materials.
6. A blister-forming agent that is an oily, colorless-to-dark-brown liquid with an odor of geraniums.
9. A nerve gas that is both a contact hazard and a vapor hazard; it operates by disabling the chemical connection between the nerves and their target organs.
10. The time period between the initial infection by an organism and the development of symptoms by a victim.
13. The process of sorting victims based on the severity of their injuries and medical needs to establish treatment and transportation priorities.
14. An infectious disease spread by the bacterium *Bacillus anthracis*; it is typically found around farms, infecting livestock.
15. An explosive or incendiary device that is fabricated in an improvised manner.
16. A chemical agent that causes severe pulmonary damage; it is a by-product of incomplete combustion.

Down

1. A highly infectious disease caused by the virus *Variola*.
2. The physical or chemical process of removing any form of contaminant from a person, an animal, an object, or the environment.
3. A device created by filling a section of pipe with an explosive material.
4. A nerve agent, principally a contact hazard; an oily liquid that can persist for several weeks.
7. Weapons whose use is intended to cause mass casualties, damage, and chaos.
8. A nerve agent that is primarily a vapor hazard.
11. A highly toxic chemical agent that attacks the circulatory system.
12. An infectious disease caused by the bacterium *Yersinia pestis*, which is commonly found on rodents.

Fire Alarms

The following real case scenarios will give you an opportunity to explore the concerns associated with terrorism awareness. Read each scenario, and then answer each question in detail.

1. It is 11:30 on a Saturday morning when your engine is dispatched to a car explosion in front of a government building near the center of your community. Upon arrival, you find a heavily damaged passenger vehicle that is on fire. The building's windows have been blown out in the front and the structural status is unknown. You observe approximately 10 people who are injured. How should you proceed?

   _____
   _____
   _____
   _____
   _____
   _____

2. Your fire department contracts with a small farming community for fire protection and EMS. The government has just announced in a press release that it has uncovered some information that agroterrorists are planning a chemical attack on the U.S. food supply. During your shift meeting, your Lieutenant reviewed the department's response guidelines to terrorist incidents. What is the fire department's role in protecting communities from terrorism?

   _____
   _____
   _____
   _____
   _____
   _____

Fire Fighter II in Action

The following scenario will give you an opportunity to apply your firefighting knowledge and your fire department SOGs to the new information you learned while studying this chapter. Research your department's SOGs and answer the assignment in detail. Compare your answers with your classmates' and discuss similarities and obvious differences between your answers.

Your community has scheduled several street events throughout the summer. The main street in town will be closed off to motor traffic several weekends, and several thousand attendees are expected at each event.

1. What outside agencies will your fire department have to interact with during preplanning as well as during the actual events?

2. What detection devices and safety equipment are available for your unified command structure to use during these events?

Fire Prevention and Public Education

Workbook Activities

The following activities have been designed to help you. Your instructor may require you to complete some or all of these activities as a regular part of your fire fighter training program. You are encouraged to complete any activity that your instructor does not assign as a way to enhance your learning in the classroom.

Chapter Review

The following exercises provide an opportunity to refresh your knowledge of this chapter. All questions in this chapter are Fire Fighter II level.

Matching

Match each of the terms in the left column to the appropriate definition in the right column.

_____ 1. Public education
_____ 2. EDITH
_____ 3. Stop, Drop, and Roll program
_____ 4. Fire prevention
_____ 5. Fire codes

A. Activities intended to prevent the outbreak of fires
B. Teaches residents how to get out of their homes safely during a fire or other emergency
C. Teaches techniques to reduce fire deaths and injuries
D. Instructs people on what to do if clothing catches fire
E. Regulations adopted to ensure a minimum level of fire safety

Multiple Choice

Read each item carefully, and then select the best response.

_____ 1. Helping people to understand how to prevent fires from occurring and teaching them how to react if a fire does occur are the goals of
  A. local governments.
  B. fire departments.
  C. public fire safety education.
  D. teachers.

_____ 2. The primary causes of fires in living room areas are electrical equipment and
  A. smoking.
  B. fireplaces.
  C. children playing with matches.
  D. heating devices.

CHAPTER 36

_____ 3. Activities that are intended to help prevent the outbreak of fires or to limit the damage if a fire does occur are referred to as
  A. fire prevention.
  B. fire codes.
  C. fire regulations.
  D. public awareness.

_____ 4. Kitchen fires are responsible for
  A. 8 percent of all residential fires.
  B. 22 percent of all residential fires.
  C. 34 percent of all residential fires.
  D. 44 percent of all residential fires.

_____ 5. The main objectives of fire prevention activities are to limit life loss, to prevent injuries, and
  A. to provide education.
  B. to minimize property damage.
  C. to provide an emergency response.
  D. to avoid regulation infraction.

_____ 6. Regulations that have been legally adopted by a government body with the authority to pass laws and enforce safety regulations are called
  A. fire bylaws.
  B. jurisdictional laws.
  C. jurisdictional regulations.
  D. fire codes.

_____ 7. Accumulated trash and a visible house number should be checked during the
  A. walk-by assessment.
  B. interior survey.
  C. exterior survey.
  D. home escape plan.

_____ 8. Which of the following is a set of documents produced by the National Fire Protection Association that is intended to address a wide range of issues relating to fire and safety?
  A. National Fire Codes
  B. National Training Standards
  C. Jurisdictional Regulations
  D. Recommended Occupant's Practices

_____ 9. A voluntary inspection of a private dwelling is called a
  A. fire department visitation.
  B. public fire safety inspection.
  C. legal requirement.
  D. home fire safety survey.

_____ 10. The process of trying to establish the cause of a fire through careful investigation and analysis of available evidence is
  A. called fire investigation.
  B. called fire cause determination.
  C. the responsibility of law enforcement agencies.
  D. regulated through fire codes.

Fundamentals of Fire Fighter Skills

Vocabulary

Define the following terms using the space provided.

1. Fire prevention:

   _____
   _____
   _____

2. Fire code:

   _____
   _____
   _____

Fill-in

Read each item carefully, and then complete the statement by filling in the missing word(s).

1. Teach students to use the back of a(n) _____ _____ to sense the temperature of a door.

2. Your highest priority as a fire fighter should always be to _____ fires.

3. The most common causes of fires in _____ are defective wiring, improper use of candles, and children playing with matches.

4. Home surveys should be conducted in a(n) _____ fashion for both the inside and the outside of the home.

5. Citizens have a(n) _____ obligation to comply with fire codes.

6. In many cases, the fire code does not apply to the _____ of a private dwelling.

7. Generally, fire codes apply to all _____, new or old.

8. Every fire fighter should know how to conduct a(n) _____ _____ safety survey in a private dwelling.

9. Fire codes are closely related to _____ codes.

10. Stress the importance of keeping _____ _____ in working order.

True /False

If you believe the statement to be more true than false, write the letter "T" in the space provided. If you believe the statement to be more false than true, write the letter "F."

_____ 1. Every kitchen should be equipped with an approved ABC-rated fire extinguisher.

_____ 2. Everyone should test their smoke alarms once a year using the test button.

_____ 3. Having working smoke alarms on each level of a house reduces the risk of death from fire by 50 percent.

_____ 4. According to the NFPA, the top five causes of fire in homes are cooking equipment, heating equipment, intentional fires, electrical distribution, and open flames.

_____ 5. After a jurisdiction adopts a fire code, it must be able to enforce that code.

_____ 6. The fire code is enforced through a legal process similar to the way traffic regulations are enforced.

_____ 7. As part of your EDITH presentation, you should stress the importance of keeping bedroom doors open during sleeping hours.

_____ 8. Residential fire safety surveys can be conducted without the occupant's permission.

_____ 9. Remind students to avoid installing smoke alarms in kitchens and garages, or near fireplaces, windows, and exit doors.

_____ 10. Good housekeeping is one of the most important issues when addressing fire safety in garages and basements.

Short Answer

Complete this section with short written answers using the space provided.

1. Identify five recommendations for kitchen safety.

_____
_____
_____
_____
_____
_____

2. List five important smoke alarm tips.

_____
_____
_____
_____
_____
_____

3. Provide six examples of public fire safety education programs.

_____
_____
_____
_____
_____
_____

4. Identify elements of the following public fire safety education programs:

A. Stop, Drop, and Roll:

_____
_____
_____
_____
_____
_____

B. Exit Drills in the Home:

_____
_____
_____
_____
_____
_____

C. Importance of smoke alarms:

_____
_____
_____
_____
_____
_____

Word Fun

The following crossword puzzle is an activity provided to reinforce correct spelling and understanding of terminology associated with firefighting. Use the clues provided to complete the puzzle. Do not include spaces or punctuation when filling in the puzzle.

Clues

Across
3. A set of legally adopted rules and regulations designed to prevent fires and protect lives in the event of a fire.
5. Change Your Clock—Change Your _____ is one example of a public fire safety education program.
6. Activities conducted to prevent fires and protect lives and property in the event of a fire.

Down
1. Device that should be installed on every floor and in or near every sleeping room of the home.
2. Fire cause _____ is the process of establishing the cause of a fire through careful investigation and analysis of available evidence.
4. A public safety program designed to teach occupants how to exit a house safely in the event of a fire or other emergency.
7. The Stop, Drop, and _____ program can be taught to children as young as preschoolers, but is also valuable for adults.

Fire Alarms

The following real case scenarios will give you an opportunity to explore the concerns associated with fire prevention and public education. Read each scenario, and then answer each question in detail.

1. You are in the middle of a home safety survey of a private dwelling. So far, the inspection is going well. You enter the basement and notice that the occupant is storing five full gasoline cans next to the gas water heater. How should you proceed?

2. Your Captain tells you that a group of 10 teenagers will visit the fire station for a tour during your next shift. He tells you to prepare an appropriate presentation and be ready to give them a tour of the station. How do you proceed?

Fire Fighter II in Action

The following scenario will give you an opportunity to apply your firefighting knowledge and your fire department SOGs to the new information you learned while studying this chapter. Research your department's SOGs and answer the assignment in detail. Compare your answers with your classmates' and discuss similarities and obvious differences between your answers.

You have been requested to do a fire safety survey at a residential single-family dwelling.

1. List the different hazards you may encounter both inside the structure and outside.

2. What fire safety education programs could you suggest during your survey?

3. What suggestions could you give the occupants to help prevent fires in their home?

Skill Drills

Skill Drills 36-3: Installing and Maintaining Smoke Alarms

Test your knowledge of this skill drill by filling in the correct words in the photo captions.

© Jones & Bartlett Learning. Photographed by Glen E. Ellman.

1. Ensure that smoke alarms are mounted on the _____ or as _____ as possible on walls.

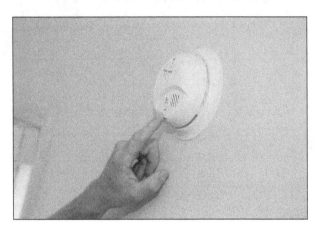
© Jones & Bartlett Learning. Photographed by Glen E. Ellman.

2. Test smoke alarms _____ a month using the _____ _____.

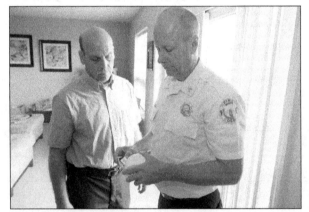
© Jones & Bartlett Learning. Photographed by Glen E. Ellman.

3. Change _____ batteries in the smoke detector every _____ months if not a 10 year battery.

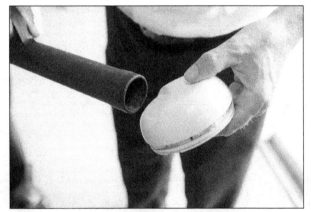
© Jones & Bartlett Learning. Photographed by Glen E. Ellman.

4. _____ alarms regularly to prevent false alarms.

Skill Drill 36-4: Home Safety Survey

Test your knowledge of this skill drill by placing the steps below in the correct order. Number the first step with a "1," the second step with a "2," and so on.

© Jones & Bartlett Learning. Photographed by Glen E. Ellman.

_____ Check inside the house for properly working smoke alarms, fire extinguishers, and fire sprinkler systems. Explain how improper cooking procedures can start kitchen fires. Stress the safe storage of cooking oils and flammable objects away from the stove. Explain the safe use of fireplaces, heating stoves, and portable heaters.

© Jones & Bartlett Learning. Photographed by Glen E. Ellman.

_____ Leave fire and life safety brochures with the building occupants.

© Jones & Bartlett Learning. Photographed by Glen E. Ellman.

_____ Look outside the house for accumulated trash, overgrown shrubs, and blocked exits.

© Jones & Bartlett Learning. Photographed by Glen E. Ellman.

_____ Review the results of the fire safety survey with the building occupants.

Fire Detection, Protection and Suppression Systems

Workbook Activities

The following activities have been designed to help you. Your instructor may require you to complete some or all of these activities as a regular part of your fire fighter training program. You are encouraged to complete any activity that your instructor does not assign as a way to enhance your learning in the classroom.

Chapter Review

The following exercises provide an opportunity to refresh your knowledge of this chapter. All questions in this chapter are Fire Fighter II level.

Matching

Match each of the terms in the left column to the appropriate definition in the right column.

_____ 1. Cross-zoned system **A.** The valve assembly on a dry sprinkler system that prevents water from entering the system until the air pressure is released

_____ 2. False alarm **B.** A sprinkler head usually marked SSU

_____ 3. Early suppression fast response sprinkler head **C.** A sprinkler system in which the pipes are normally filled with water

_____ 4. OS&Y valve **D.** A sprinkler head designed to react quickly and suppress a fire in its early stages

_____ 5. Dry-pipe valve **E.** A device that increases the removal of the air from a dry-pipe or preaction sprinkler system

_____ 6. Upright sprinkler head **F.** A sprinkler control valve with a valve stem that moves in and out as the valve is opened or closed

_____ 7. Gas detector **G.** The activation of a fire alarm system when there is no fire or emergency condition

_____ 8. Accelerator **H.** A fire alarm system that requires activation of two separate detection devices before initiating an alarm

_____ 9. Nuisance alarm **I.** A device that measures the concentration of dangerous gases

_____ 10. Wet sprinkler system **J.** A device within a piping system that allows water to flow in only one direction

_____ 11. Line detector **K.** A standard fire alarm audible signal for alerting occupants of a building

_____ 12. Clapper mechanism **L.** Wire or tubing that can be strung along the ceiling of large open areas to detect an increase in heat

_____ 13. Temporal-3 pattern **M.** A fire alarm signal caused by malfunction or improper operation of a fire alarm system or component

_____ 14. Deluge head **N.** A sprinkler head that has no release mechanism

CHAPTER 37

Multiple Choice

Read each item carefully, and then select the best response.

_____ 1. Which of the following is the name for a special extinguishing system that operates by discharging a gaseous agent into the atmosphere at a concentration that will extinguish a fire?
   A. Wet chemical extinguishing system
   B. Clean agent extinguishing system
   C. Dry chemical extinguishing system
   D. Carbon dioxide extinguishing system

_____ 2. Which type of sprinkler head has a glass bulb filled with glycerin to hold the cap in place?
   A. Fusible-link sprinkler head
   B. Chemical-pellet sprinkler head
   C. Deluge head
   D. Frangible-bulb sprinkler head

_____ 3. The network of pipes and outlets for fire hoses built into a structure and designed for use by the building occupants is designated as belonging to
   A. Class I.
   B. Class II.
   C. Class III.
   D. Class IV.

_____ 4. The most current codes require new homes to have a smoke alarm
   A. on every floor.
   B. in every room.
   C. in every bedroom and on every floor level.
   D. in every bedroom, hallway, and floor level.

_____ 5. To allow the fire department's engine to pump water into the sprinkler system, each sprinkler system should also have a
   A. primary feeder.
   B. secondary feeder.
   C. pumper outlet.
   D. fire department connection.

_____ 6. A network of pipes and outlets for fire hoses built into a structure to provide water for firefighting purposes is called a
   A. residential pipe system.
   B. grid system.
   C. standpipe system.
   D. closed flow system.

_____ 7. Fire alarm systems are activated by the
   A. remote annunciator.
   B. ESFR device.
   C. alarm initiation device.
   D. line detector.

_____ 8. Which type of valve is mounted on the outside wall of a building?
   A. PIV
   B. OS&Y valve
   C. WPIV
   D. Support control valve

_____ 9. In most cases, the entire sprinkler system can be shut down by
   A. closing the main control valve.
   B. using the remote annunciator panel.
   C. deactivating the alarm.
   D. using a sprinkler wedge.

_____ 10. Which type of detectors are triggered by the invisible products of combustion?
   A. Ionization smoke detectors
   B. Photoelectric smoke detectors
   C. Heat detectors
   D. Spot detectors

_____ 11. Which type of detectors detect the electromagnetic light waves produced by a flame?
   A. Beam detectors
   B. Line detectors
   C. Air sampling detectors
   D. Flame detectors

_____ 12. Most modern sprinkler systems are connected to the building's fire alarm system to alert the occupants by a pressure switch or a
   A. tamper switch.
   B. flow switch.
   C. clapper switch.
   D. trigger switch.

_____ 13. Which type of sprinkler head is triggered by the melting of a metal alloy at a specific temperature?
   A. Fusible-link sprinkler head
   B. Frangible-bulb sprinkler head
   C. ESFR sprinkler head
   D. Pendant sprinkler head

_____ 14. Which type of fire alarm requires two steps before the alarm will activate?
   A. Single-action pull-station
   B. Double-action pull-station
   C. Protected pull-station
   D. Tamper alarm

_____ 15. The temporal-3 pattern is a(n)
   A. verification system.
   B. standardized audio pattern.
   C. alarm activation system.
   D. photoelectric detector system.

_____ 16. A smoke detector is designed to sense the presence of
   A. smoke.
   B. heat.
   C. fire.
   D. toxic gases.

_____ 17. The activation of a single smoke detector plus the activation of a second smoke detector is characteristic of a
   A. double-pull alarm system.
   B. verification system.
   C. cross-zoned system.
   D. nuisance system.

_____ 18. Which type of detectors are calibrated to detect the presence of a specific gas that is created by combustion?
   A. Gas detectors
   B. Combustion detectors
   C. Beam detectors
   D. Rate-of-calibration detectors

_____ 19. Which type of detectors use wire or tubing strung along the ceiling of large open areas to detect an increase in heat?
   A. Spot detectors
   B. Heat detectors
   C. Beam detectors
   D. Line detectors

_____ 20. Many buildings have an additional fire alarm control display panel in the front of the building called a
   A. remote alarm station.
   B. remote control panel.
   C. remote annunciator.
   D. remote visual board.

Vocabulary

Define the following terms using the space provided.

1. Zoned system:

2. Deluge sprinkler system:

3. Verification system:

4. Outside stem and yoke valve:

Fundamentals of Fire Fighter Skills

5. Post indicator valve:
   _____
   _____
   _____

6. Fire department connection:
   _____
   _____
   _____

Fill-in
Read each item carefully, and then complete the statement by filling in the missing word(s).

1. A(n) _____ alarm occurs when an alarm system is activated by a condition that is not really an emergency.

2. The Class _____ standpipe is designed for use by fire department personnel only.

3. A(n) _____ detector is a type of photoelectric smoke detector used to protect large open areas.

4. The network of pipes that delivers water through the sprinkler system is the sprinkler _____.

5. A(n) _____ _____ system sends a signal directly to the fire department or to a monitoring location via telephone or radio signal.

6. _____ sprinkler heads are designed for horizontal mounting and projecting out from a wall.

7. The fire alarm control panel serves as the _____ of the fire alarm system.

8. _____ _____ _____ heat detectors will be activated if the temperature of the surrounding air rises more than a set amount in a given time period.

9. The fire alarm control panel should monitor the entire alarm system to detect any _____.

10. _____ detectors are triggered by the visible products of combustion.

True/False

If you believe the statement to be more true than false, write the letter "T" in the space provided. If you believe the statement to be more false than true, write the letter "F."

_____ 1. An activated alarm sounds throughout a building.

_____ 2. Smoke alarms can be either hard-wired to a 110-volt electrical system or battery operated.

_____ 3. All sprinkler systems should be equipped with a method for sounding an alarm whenever there is water flowing in the pipes.

_____ 4. Fire alarm systems can control doors and elevators.

_____ 5. An activated sprinkler head in an automatic sprinkler system triggers the water-motor gong.

_____ 6. A photoelectric detector has a small amount of radioactive material inside a chamber.

_____ 7. A central station is operated by the fire department.

_____ 8. Bimetallic strips are made to respond to a rapid increase in temperature.

_____ 9. Heat detectors provide reliable life-safety protection.

_____ 10. Nuisance alarms are caused by individuals who deliberately activate fire alarms when there is no fire.

Short Answer

Complete this section with short written answers using the space provided.

1. List the five fire department notification systems.

   _____
   _____
   _____
   _____
   _____

2. Identify the three fire suppression systems.

   _____
   _____
   _____
   _____
   _____

3. Identify the four categories of sprinkler systems and provide a brief description of each category.

   _____
   _____
   _____
   _____
   _____

4. Identify the three categories of standpipes, with a description of their intended use.

5. List the three basic components of a fire alarm system, and describe their functions.

6. Identify the four types of sprinkler heads.

7. Identify the different styles of indicator valves.

Word Fun

The following crossword puzzle is an activity provided to reinforce correct spelling and understanding of terminology associated with firefighting. Use the clues provided to complete the puzzle. Do not include spaces or punctuation when filling in the puzzle.

CLUES

Across

1. A fire hose connection through which the fire department can pump water into a sprinkler system or standpipe system.
3. A liquefied gas extinguishing agent that puts out a fire by chemically interrupting the combustion reaction between fuel and oxygen. These agents leave no residue.
5. A sprinkler control valve with a valve stem that moves in and out as the valve is opened or closed.
8. A standpipe system designed for use by fire department personnel only. Each outlet should have a valve to control the flow of water and a 2½-inch (64-mm) male coupling for a fire hose.
9. A standpipe system designed for use by occupants of a building only. Each outlet is generally equipped with a length of 1½-inch (38-mm) single-jacket hose and a nozzle, which are preconnected to the system.
10. A sprinkler control valve with an indicator that reads either open or shut depending on its position.
11. A fire alarm system design that divides a building or facility into zones and has audible notification devices that can be used to identify the area where an alarm originated.
12. A valve assembly designed to release water into a sprinkler system when an external initiation device is activated.
13. A mechanical device installed within a piping system that allows water to flow in only one direction.

Down

1. A sensing device that detects the radiant energy emitted by a flame.
2. A fire alarm system design that divides a building or facility into zones so that the area where an alarm originated can be identified.
4. An alarm system that provides no information at the alarm control panel indicating where the activated alarm is located.
6. The valve assembly on a dry sprinkler system that prevents water from entering the system until the air pressure is released.
7. A valve that signals an alarm when a sprinkler head is activated and prevents nuisance alarms caused by pressure variations.

Fire Alarms

The following real case scenarios will give you an opportunity to explore the concerns associated with fire protection, suppression, and detection systems. Read each scenario, and then answer each question in detail.

1. It is 10:00 on a Wednesday morning when your engine is dispatched to deal with an alarm activation at an office building. You and the crew check the annunciator panel. The annunciator indicates that the carbon dioxide extinguishing system has been activated in the computer server room. One of the managers of the company meets you at the door and reports that light smoke was seen in the server room. What are the special hazards of a carbon dioxide extinguishing system?

___

2. It is Sunday morning and you have completed checking the apparatus. Your Lieutenant calls the crew together for a practice drill. She tells you that the drill will familiarize you with the fire suppression system at the new city courthouse. The fire suppression system at the courthouse has a supplied wet sprinkler system with fire department connections. Why is it important for fire fighters to have a basic understanding of fire suppression systems?

___

Fire Fighter II in Action

The following scenarios will give you an opportunity to apply your firefighting knowledge and your fire department SOGs to the new information you learned while studying this chapter. Research your department's SOGs and answer the assignments in detail. Compare your answers with your classmates' and discuss similarities and obvious differences between your answers.

Most fire fighters respond to and fight fires in both residential and commercial/industrial buildings. Many of these structures have sprinkler systems, but there are differences between residential sprinkler systems and those systems used in commercial and industrial structures.

1. What are these differences, and how will these differences affect our response?

   _____
   _____
   _____
   _____
   _____

Most fire departments have high-rise buildings or tall structures such as silos and grain elevators in their jurisdiction.

2. Identify some of the problems you may have when using a standpipe system in these tall buildings.

   _____
   _____
   _____
   _____
   _____

3. How can you minimize or eliminate these problems?

   _____
   _____
   _____
   _____
   _____

Fire Cause Determination

Workbook Activities

The following activities have been designed to help you. Your instructor may require you to complete some or all of these activities as a regular part of your fire fighter training program. You are encouraged to complete any activity that your instructor does not assign as a way to enhance your learning in the classroom.

Chapter Review

The following exercises provide an opportunity to refresh your knowledge of this chapter.

Matching

Match each of the terms in the left column to the appropriate definition in the right column.

_____ 1. Ignitable liquid **A.** Intentionally set fires

_____ 2. Undetermined **B.** Fire cause classification that includes fires for which the cause has not been or cannot be proven

_____ 3. Contaminated **C.** Evidence that is reported first-hand

_____ 4. Incendiary fires **D.** A pathological fire-setter

_____ 5. Pyromaniac **E.** Items that can be examined in a laboratory and presented in court to prove or demonstrate a point

_____ 6. Arsonist **F.** Classification for liquid fuels including both flammable and combustible liquids

_____ 7. Circumstantial evidence **G.** A term used to describe evidence that may have been altered from its original state

_____ 8. Trace evidence **H.** A person who deliberately sets a fire to destroy property with criminal intent

_____ 9. Direct evidence **I.** The means by which alleged facts are proven by deduction or inference from other facts that were observed first hand

_____ 10. Physical evidence **J.** Evidence of a minute quantity that is conveyed from one place to another

CHAPTER 38

Multiple Choice

Read each item carefully, and then select the best response.

_____ 1. Most fires, fire deaths, and injuries occur in
   A. residential occupancies.
   B. industrial settings.
   C. small businesses.
   D. gas or fueling stations.

_____ 2. If a fire intensifies in a short period of time, it may indicate
   A. poor dispatch information.
   B. the use of an accelerant.
   C. extreme weather conditions.
   D. multiple points of origin.

_____ 3. What might charring on the underside of a low horizontal surface, such as a tabletop, indicate?
   A. The fire's point of origin was on top of the surface.
   B. The fire was accidental.
   C. There was a pool of a flammable liquid.
   D. The fire started from a cigarette butt.

_____ 4. An inverted cone pattern on a wall indicates that
   A. a flammable liquid was used to start the fire.
   B. there was a wide point of origin for the fire.
   C. the fire began on the ceiling.
   D. there was a flashover.

_____ 5. An accelerant may have been used if the fire _____ when water is applied.
   A. spreads
   B. grows
   C. is extinguished
   D. rekindles

_____ 6. The cause of a fire can be classified as either incendiary or
   A. malicious intent.
   B. criminal intent.
   C. undetermined.
   D. accidental.

_____ 7. Which of the following terms is used to describe a device or mechanism that is used to start a fire?
   A. Arsonist device
   B. Incendiary device
   C. Accelerant
   D. Trailer

_____ 8. In most jurisdictions, the _____ determines the cause of a fire.
   A. law enforcement agency
   B. chief of the fire department
   C. property owner
   D. fire investigator

_____ 9. A charred V-pattern on a wall indicates that fire spread
   A. along the floor before reaching the wall.
   B. up and out from an unknown material at the base of the V.
   C. along the ceiling before reaching the wall.
   D. slowly.

_____ 10. "Chain of custody" is a legal term that describes the process of maintaining continuous possession and control of evidence from
   A. the time it is discovered until it is presented in court.
   B. investigator to incident commander.
   C. investigator to law enforcement agency.
   D. discovery to isolation.

_____ 11. One of the first steps in a fire investigation is identifying the
   A. fuel supply.
   B. accelerants.
   C. point of origin.
   D. evidence.

_____ 12. An arson investigation must determine not only who was responsible for starting the fire, but also
   A. why the person started the fire.
   B. when the person started the fire.
   C. the property damage caused by the fire.
   D. the cause and origin of the fire.

_____ 13. A pyromaniac is a pathological fire-setter who is often a(n)
   A. juvenile male.
   B. adolescent female.
   C. adult female.
   D. adult male.

_____ 14. The process of carefully looking for evidence within the debris is referred to as
   A. layering.
   B. overhaul.
   C. evidence recovery.
   D. digging out.

_____ 15. An arsonist who sets three or more fires at separate locations with no emotional cooling-off period between fires is called a
   A. spree arsonist.
   B. serial arsonist.
   C. mass arsonist.
   D. motivated arsonist.

_____ 16. An arsonist who sets three or more fires, with a cooling-off period between fires, is called a
   A. spree arsonist.
   B. serial arsonist.
   C. mass arsonist.
   D. motivated arsonist.

_____ 17. Preadolescent male fire-starters are most often motivated to start fires by
   A. a need for attention.
   B. curiosity.
   C. revenge or spite.
   D. excitement.

___ **18.** The fire department's authority over an incident ends when
   A. the property is formally released to the property owner.
   B. the property is under the investigator's supervision.
   C. any criminal or malicious intent regarding the fire's origin is ruled out.
   D. the property is secured and no hazards to public safety exist.

___ **19.** Which type of evidence can be used to prove a theory, based on facts that were observed first-hand?
   A. Demonstrative evidence
   B. Trace evidence
   C. Circumstantial evidence
   D. Physical evidence

___ **20.** An arsonist who sets three or more fires at the same site or location during a limited period of time is called a
   A. spree arsonist.
   B. serial arsonist.
   C. mass arsonist.
   D. motivated arsonist.

Vocabulary

Define the following terms using the space provided.

1. Competent ignition source:

   _____
   _____
   _____

2. Chain of custody:

   _____
   _____
   _____

3. Trailers:

   _____
   _____
   _____

4. Depth of char:

   _____
   _____
   _____

Fill-in
Read each item carefully, and then complete the statement by filling in the missing word(s).

1. An arsonist may place _____ to hinder the efforts of fire fighters.

2. _____ removed from any victim should be preserved as evidence.

3. The _____ of any victims found in the building should be noted.

4. Smoke residue and _____ patterns can be helpful in identifying the point of origin.

5. Evidence should not be _____ or altered from its original state in any way.

6. What a fire fighter _____ during an incident could be significant in an investigation of an incident.

7. At the point of origin, an ignition source comes into contact with a(n) _____ _____.

8. Until the fire is under control, fire fighters must concentrate on fighting the fire and not investigating the _____.

9. The fire investigation process usually begins with an examination of the building's _____.

10. Anything that can be used to validate a theory is _____ evidence.

11. Insurance companies often investigate fires to determine the _____ of a claim.

12. A fire investigation can provide many _____, even if the specific cause and origin are never determined.

True/False
If you believe the statement to be more true than false, write the letter "T" in the space provided. If you believe the statement to be more false than true, write the letter "F."

_____ 1. The appearance and behavior of people at the scene of a fire can provide valuable clues.

_____ 2. A cause-and-origin investigation determines where, why, and how a fire originated.

_____ 3. Evidence is most often found during the size-up phase of a fire.

_____ 4. The most important reason for investigating accidental fires is to identify the causes.

_____ 5. Arsonists often open the shades and windows of structures they burn.

_____ 6. Charring is usually deepest on the edges of the object.

_____ 7. The color of the smoke often indicates what is burning.

_____ 8. A fire can be caused by an act or by an omission.

_____ 9. To avoid contaminating evidence, fire investigators always wash their tools between taking samples.

_____ 10. Only one person should be responsible for collecting and taking custody of all evidence at a fire scene.

_____ 11. For most fire fighters, "The fire is under investigation" is the best reply to any questions concerning the cause of the fire.

_____ 12. To assist with evidence collection, burned materials should be thrown into a pile.

Short Answer

Complete this section with short written answers using the space provided.

1. Describe the characteristics of fires set by pyromaniacs.

2. List the six common arson motives listed in NFPA 921, *Guide to Fire and Explosion Investigations*.

3. Describe the steps to take if a fire investigator is not available and the premises needs to be maintained under the control of the fire department until the investigation can take place.

4. Describe the role and relationship of the fire fighter II to criminal investigators and insurance investigators.

5. Describe how the origin and cause of a fire are determined.
   _____
   _____
   _____
   _____
   _____
   _____
   _____

6. Describe how to assist the fire investigators in the process of digging out the fire scene.
   _____
   _____
   _____
   _____
   _____
   _____
   _____

7. Describe techniques for preserving fire-cause evidence.
   _____
   _____
   _____
   _____
   _____
   _____
   _____

Word Fun

The following crossword puzzle is an activity provided to reinforce correct spelling and understanding of terminology associated with firefighting. Use the clues provided to complete the puzzle. Do not include spaces or punctuation when filling in the puzzle.

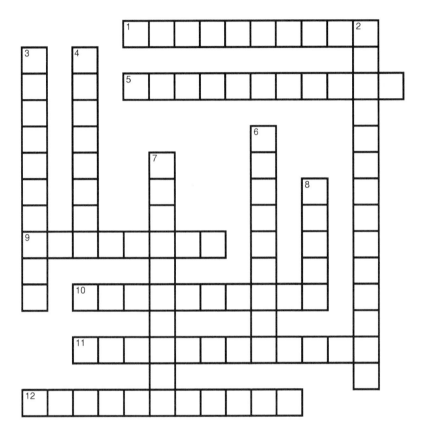

CLUES

Across

1. A pathological fire-setter.
5. The thickness of the layer of a material that has been consumed by a fire. On wood, this can be used to help determine the intensity of a fire at a specific location.
9. A person who deliberately sets a fire to destroy property with criminal intent.
10. A series of fires started by an arsonist who sets three or more fires at separate locations, with no emotional cooling-off period between fires.
11. A term used to describe evidence that may have been altered from its original state.
12. A series of fires set by the same offender, with a cooling-off period between fires.

Down

2. A legal term used to describe the paperwork or documentation describing the movement, storage, and keeping of evidence, such as the record of possession of a gas can from a fire scene.
3. A fire that is intentionally ignited under circumstances in which the person knows that the fire should not be ignited.
4. Combustible materials (such as rolled rags, blankets, and newspapers or ignitable liquid) that are used to spread fire from one point or area to other points or areas, often used in conjunction with an incendiary device.
6. Arson in which an offender sets three or more fires at the same site or location during a limited period of time.
7. Fire-cause classification that includes fires with a proven cause that does not involve a deliberate human act.
8. The crime of maliciously and intentionally, or recklessly, starting a fire or causing an explosion. (NFPA 921)

Fire Alarms

The following real case scenarios will give you an opportunity to explore the concerns associated with fire cause determination. Read each scenario, and then answer each question in detail.

1. Your engine has responded to a structure fire in a church. A serial arsonist has been setting fire to churches in your community. This fire has some of the same characteristics as the other church fires. You and your partner confined the fire to the basement of the church. Now you and your partner have been assigned to help the fire investigator overhaul the fire scene. How should you proceed?

   _____
   _____
   _____
   _____
   _____
   _____

2. You are on the scene of a suspected arson at a waterfront restaurant. You are assigned to work with one of the arson investigators in your department to help collect and process possible evidence. How do you proceed?

   _____
   _____
   _____
   _____
   _____
   _____

🔥 Fire Fighter II in Action

The following scenario will give you an opportunity to apply your firefighting knowledge and your fire department SOGs to the new information you learned while studying this chapter. Research your department's SOGs and answer the assignment in detail. Compare your answers with your classmates' and discuss similarities and obvious differences between your answers.

During the overhaul stage of a particularly well-involved house fire, you notice one company being very aggressive in their work and beginning to turn the overhaul into a company competition. The cause of the fire has not been determined; in fact, the state investigator has not arrived yet.

1. Is there a cause for concern?

   _____
   _____
   _____
   _____
   _____

2. If so, how should you handle this?

   _____
   _____
   _____
   _____
   _____

CPSIA information can be obtained
at www.ICGtesting.com
Printed in the USA
JSHW012025261219
3201JS00002B/6